Eyewitnesses to Peace

Letters from Canadian Peacekeepers

by

Jane Snailham

edited by Alex Morrison and Steven A. Torrisi

Modern international stability operations frequently involve several warring factions, an unstable or non-existent truce, and a national theatre of operations. To deal with these operations there is a *New Peacekeeping Partnership:* The ***New Peacekeeping Partnership*** is the term applied to those organizations and individuals that work together to improve the effectiveness of modern peacekeeping operations. It includes the military; civil police; government and non-government agencies dealing with human rights and humanitarian assistance; diplomats; the media; and organizations sponsoring development and democratization programmes. The Pearson Peacekeeping Centre serves the *New Peacekeeping Partnership* by providing national and international participants with the opportunity to examine specific peacekeeping issues, and to update their knowledge of the latest peacekeeping practices.

Canadian Cataloguing in Publication Data

Snailham, Jane, 1941-

Eyewitnesses to peace

ISBN 1-896551-16-5

1. Canada -- Armed Forces -- Foreign countries. 2. United Nations -- Canada. 3. United Nations -- Armed Forces. I. Morrison, Alex, 1941- II. Torrisi, Steven A. III. Title.

JX1981.P7S62 1998   355.3'57'092   C98-950212-0

Front cover photo: Memorial to Canadian Peacekeepers, Ottawa.
Credit: Armand F. Wigglesworth.

The editors and publisher thank the United Nations for permission to base the maps contained in this book on UN maps.

Printed by Brown Book Company Ltd., Toronto, ON

Eyewitnesses to Peace

Letters from Canadian Peacekeepers

by

Jane Snailham

edited by Alex Morrison and Steven A. Torrisi

The Canadian Peacekeeping Press
1998

**The Lester B. Pearson
Canadian International Peacekeeping
Training Centre**
President, Alex Morrison, MSC, CD, MA

The Pearson Peacekeeping Centre supports and enhances the Canadian contribution to international peace, security, and stability. The Centre conducts research and provides advanced training and educational programmes, and is a division of the Canadian Institute of Strategic Studies. The Canadian Peacekeeping Press is the publishing division of the Pearson Peacekeeping Centre.

Canadian Peacekeeping Press publications include:

Peacekeeping at a Crossroads (1997)

Refugees, Resources and Resoluteness (1997)

Multilateralism and Regional Security (1997)

Theory, Doctrine and Practice
of Conflict De-Escalation in Peacekeeping Operations (1997)

Facing the Future:
Proceedings of the 1996 Canada-Japan Conference on Modern Peacekeeping
(1997)

Peacekeeping with Muscle (1997)

Seeds of Freedom:
Personal Reflections on the Dawning of Democracy (1996)

Analytic Approaches to the Study of Future Conflict (1996)

The Centre-Periphery Debate in International Security (1996)

Rapid Reaction Capabilities: Requirements and Prospects
*Les capacités de réaction rapide de l'ONU:
exigences et perspectives* (1996)

The New Peacekeeping Partnership (1995)

For publications information, please contact:

Sue Armstrong, Publications Manager
The Pearson Peacekeeping Centre
Cornwallis Park, PO Box 100
Clementsport, NS B0S 1E0 CANADA
Tel: (902) 638-8611 ext. 161 Fax: (902) 638-8576
E-mail: sarmstro@ppc.cdnpeacekeeping.ns.ca

Or visit the Pearson Peacekeeping Centre website:
http://www.cdnpeacekeeping.ns.ca

The Centre (a division of the Canadian Institute of Strategic Studies), established by the Government of Canada in 1994, is funded, in part, by the Department of Foreign Affairs and International Trade and the Department of National Defence of Canada.

Le centre (une division de l'Institut canadien d'études stratégiques) à été établi par le Gouvernement du Canada en 1994. Le soutien financier de Centre provient, en partie, des ministères des Affaires étrangères et du commerce international et de la Défense nationale.

Table of Contents

List of Photographs

List of Maps

A Prayer for Peacekeepers

Oh Heavenly Father, I pray to you

For all my peacekeeping friends —

In troubled lands, around the world

Striving to bring a peace.

The UN asks impossible tasks,

And Canadians always come through —

God grant them the courage, strength and love —

Your truth, will be their guide.

Jane Snailham

Dedication

This book is dedicated to all Canadian Peacekeepers.

A special thank you to my "peacekeeping friends" who shared their lives with me during their United Nations tours: Sheldon Alley, Gordon Anderson, Rod Babiuk, Mike Bonin, Brett Boudreau, Cecil Bouter, Carol Ann Boutilier, Kathy Branston, Gregory Burke, Heather Breton, Lynda Busby, Jame Cade, Stephen Carr, Mike Campbell, Paul Chura, Stephanie Conley, Alinah Cruz, Jill Davidson, W.M. Derousie, Tim Dunne, Douglas Drysdale, Robert Emmett, Rod Escayola, Shawn Fortin, Barry Frewer, Harold Gale, Gaby Gauthier, Christian Girard, Paul Girault, Joe Goetz, T.J. Grant, Alain Horth, Wayne Johnson, Eva Kelly, Tim Kirk, Kevin Laing, Bryan MacFarlane, Mike Manseau, Danny Macphee, Tobie Mangione, Bob Martell, Steve Martin, Steve McClusky, Kevin McLeod, G.D. McNally, David Moore, Jim Muise, D. Nearing, Len Nearing, Ken Nunn, Kent Page, Susan Paul, C.A. Peck, Rick Powell, Robert Primmer, Hal Pugh, Steve Quilty, Paul Rutherford, Clive Samson, Kevin Sampson, Shelley Spyksma, Ian Stock, Patrick Twomey, Lori Turner, Alain Vachon, Kathy Wallocha, Perry Wells, C.A. Wilson and Alison Zawada.

You are in my thoughts and prayers.

Acknowledgments

My Dad, Colonel Armand F. Wigglesworth, CM, SBro.StJ, BEM, CD, KLJ — you took the time and had the patience while editing your second book to encourage me, to give me much needed help and to trust me with your computer! I appreciate your love and support.

My sister Joanne, who during a most difficult period in life, gave me all her love and encouragement.

Major Douglas Martin, without our chats and your support, I would not be writing to our peacekeepers today.

Sergeant Mike Bonin, through your slides I learned the real impact of the conflict in the Balkans. Thank you for sharing, and being patient with my questions.

Major Tim Dunne, in 1992, you gave me the address that enabled me to write to "my first peacekeeper" and your support, thank you.

Captain Ian Stock and Wendy Clark, thank you for always being there for me.

Lieutenant-Commander Douglas Drysdale, thank you for your help and support.

Steven A. Torrisi, thank you for your hard work and dedication to this book and for your help in making a dream come true.

Foreword

There are many great Canadians who are well known. On the other hand, the public knows little about the brotherhood of Canadians who wear the "Blue Beret" of the United Nations. If not for the selfless efforts of Jane Snailham, a resident of Nova Scotia, the public would never have heard their stories and emotions about service for Canada, the world and peace.

Her personal quest to support Canadians serving overseas in UN peacekeeping operations began when she saw a television news broadcast which showed a soldier of November Company of the Royal Canadian Regiment, serving in the former Yugoslavia. Touched by the work he was doing, she decided to write him a letter of encouragement. That letter drew a response of friendship from the young infantryman. Their mutual correspondence remains in place today — long after his return to Canada.

This book was inspired by the letters Jane received from servicemen and servicewomen in every branch of the Canadian Forces. Her wish is that all Canadians read the words and feel the emotions of these letters. Jane has, however, deleted the names of the letter writers because the letters are personal and wants those who continue to write to her to know that she considers these to be letters of confidence.

Without detracting from the respect and recognition Jane bestowed upon these unsung heroes — I believe she is a true hero. It is Canadians such as her who make us proud to don our uniforms and carry the "Red Maple Leaf" on behalf of Canada to war-torn countries around the world.

— Major Douglas Martin,
Army Public Affairs,
Edmonton

About the Author

Jane Snailham, a former nurse, craftsperson, wife and mother is a champion of Canada's Armed Forces, and of Canada's peacekeepers. Since 1992, she has been sending her pride, encouragement, and support to Canadian peacekeepers around the world. Jane accompanied most of her letters with a small gold guardian angel pin. Jane lives in Halifax, with her husband Bill and son Scott.

Author Jane Snailham shown here with photos of many of the Canadian Peacekeepers to whom she has written.

Preface

Imagine being away from home, for months at a time, facing the unknown. Imagine seeing your partner's head grazed by a sniper's bullet. Imagine escorting food convoys routinely targeted by mines, mortars and snipers. Imagine digging graves for victims of a massacre, and then having to bury them.

It is difficult for us in Canada to comprehend any of these situations; but, this is all in a day's work for Canadian peacekeepers. Canadian troops have donned the Blue Beret in almost every peacekeeping and military observer mission. The responsibility bestowed upon an individual wearing the Blue Beret includes being away from home for long periods, encountering highly volatile situations and working 12 to 14 hours a day, seven days a week.

As our world changes and more people seek the universal right to peace, order and good government Canadians are not just "referees" on a separation line between opposing sides. Along with keeping the peace, they are rebuilding and restoring peace, "all in the line of duty." They have helped re-settle refugees, train law-abiding constabularies, teach English and French, restore public works, supervise elections, monitor illegal smuggling and oversee administrative services in order to return war-torn nations to some semblance of normalcy.

With an international reputation of tolerance, patience, compassion and expertise, Canadians can truly be proud of the contributions our armed forces have made serving in the field. Many have paid the supreme sacrifice, a reminder to us all that peacekeeping does not always mean peace. Through all this, pride in Canada, pride in their mission and pride in themselves shines brightly. They are making a difference!

Since March 1992, I have had the privilege of writing to, and receiving letters from our Canadian peacekeepers on several UN missions. I have laughed and cried with them, learned the joys they feel, their frustrations, their accomplishments; and, I have felt their loneliness.

I have been asked many times why and what do I write to someone I do not know. When I wrote a letter to my first peacekeeper, a fellow Nova Scotian, I just had to let him know I was very proud of him! In my letters, I introduce myself and my family and I discuss daily life back in Canada. I ask them about their interests, and they tell me whatever they would like to tell me. Our soldiers are surprised to receive letters from a stranger as they only expect them from family and friends. It has truly been a learning and rewarding experience for me, and I consider it an honour and a privilege to write these brave Canadians.

As you read these letters, you will understand a peacekeeper is not just a soldier, but a caring human being. These letters are their stories. It is my hope that with understanding, will come appreciation and pride.

Jane Snailham
Halifax, Nova Scotia
August 1998

Introduction

During the past few years I have been sent a number of manuscripts for comment consisting of collections of letters from soldiers who fought in WWI or II. The manuscripts were usually produced by the letter writer's immediate offspring or a grandchild. I never failed to be amazed that in the middle of chaos, excitement and fear, soldiers, sailors and airmen would find the time to record their thoughts in such articulate and descriptive terms. The "Big Picture" is rarely as interesting as the personal experiences of a few who are actually recording history, albeit in a very personal way, as it is being made.

Canadians serving abroad on peace operations since 1991 face a major new challenge-frustration. Our veterans from WWI, WWII and Korea faced much greater danger in most instances, however fortunately, they didn't have to watch the so called "experts" analyse their mission on prime time TV before it started nor did they have to listen to reporters describe incidents in real time to the world -wide audience knowing that there was much more to the story than was being reported. Our brave veterans of Vimy Ridge, Normandy and Kapyong didn't have to tolerate some commentators referring to their undertaking as "Foolhardy", "Not in the national interest", "A potential slaughter", "Politically motivated" etc, before they fired the first shots. Blessedly, they did their job (and excelled) before the public and the pundits were brought into the picture.

The modern peacekeeper is frequently bombarded with commentary that questions the wisdom of his or her mission. When the UN: Security Council passes a resolution authorizing a particular peacekeeping mission it is usually the result of compromise between the five permanent members: USA, Russia, China, France, And Britain. Direction is rarely, if ever, clear at the top and therefore it is always less clear when it gets to the soldier on the ground in the mission area. This fact combined with frequent negative media comments from home and International Community elevates the level of frustration experienced by many peacekeepers.

Fortunately, soldiers know how to cope with this aspect of peacekeeping operations. They take their satisfaction and it's considerable, from day to day contact with those they are helping. The impact of an ill-informed pundit questioning the wisdom of a mission is easily off set in a peacekeepers mind by the overwhelming response of a single individual when food and medicine is delivered in the midst of chaos or when a population exchange is brokered between warring factors thereby reuniting family members. For the peacekeepers these are the measures of success and satisfaction not the grand geo political objectives of the mission.

In the midst of this frequently frustrating service I can only imagine that putting one's thoughts on paper in the form of a letter or diary must be therapeutic. In spite of the advent of more frequent access to telephones and e mail, "snail mail", with its slower pace and intimacy allows the writer more time to paint a picture in words, more time to express thoughts that are all to often overtaken by the urgency associated with more modern communications.

The letters received by Jane and published here are more descriptive of a service abroad than any news report or film footage could ever hope to show. These are the thoughts and opinions of our front line ambassadors in uniform, They are respected without equal, around the world. In the absence of scandal they labour out of the limelight doing the dirty work for the rest of us. Bless them every one and bless Jane Snailham for sharing their thoughts with us.

Lewis W. MacKenzie MSC,CD
Major General (retd)

The Former Yugoslavia

The Canadian Contribution to the Balkans

Mission Background

Although one can trace the roots of the Yugoslav Wars of Succession back hundreds of years, fighting broke out in Yugoslavia in the summer of 1991 following the decision of Slovenia and Croatia to declare their independence from the Yugoslav Federation. From 1991 to 1998, the nature of the conflict and the international community's response evolved to encompass many elements contained under the peacekeeping umbrella; peacekeeping, peace enforcement and peacebuilding being just a few. However, a lasting peace has yet to be built despite the myriad of international missions that have come and gone.

European Community Monitor Mission (ECMM)

The origins of Canada's military involvement in the Balkans pre-date the formal UN presence in the former Yugoslavia. From September 1991 (until withdrawn in March 1994), 15 CF personnel served with the ECMM under Operation *Bolster* in Albania, Bulgaria, Hungary and the former Yugoslavia. Primary tasks assigned to CF personnel included: monitoring and reporting upon the elements of cease-fire agreements, humanitarian assistance and confidence-building measures. Unlike their future UN counterparts, whose operations were restricted to particular areas, the ECMM was able to operate throughout all of the former republics of Yugoslavia.

United Nations Protection Force (UNPROFOR)

In January 1992, CF personnel seconded to the UN liaison mission deployed to the former Yugoslavia to survey the potential for deploying, sustaining and operating a peacekeeping mission. Less than a month after the establishment of UNPROFOR in February 1992, Canada authorized Operation *Harmony*, which deployed a 1,200-strong battle group (Canadian Battalion One or CANBATT One) to the region of Croatia known as Western Slavonia. The nucleus of this initial battle group was the 1st Battalion, Royal 22nd Regiment (IR22eR) and elements of the 3rd Battalion, RCR

(3RCR) and 4th Canadian Engineer Regiment (4CER). In July 1992, elements of the 1st Battalion, R22eR temporarily deployed to Bosnia-Herzegovina as an interim measure to open Sarajevo International Airport and provide escorts for the delivery of humanitarian supplies into the city. Shortly thereafter, the battle group re-deployed to Croatia after being relieved by French, Egyptian and Ukrainian battalions.

In October 1992, the Canadian government authorized the deployment of a second battle group (Canadian Battalion Two or CANBATT Two) under Operation *Cavalier* to central Bosnia-Herzegovina in order to provide convoy security for UN humanitarian assistance efforts. Local disagreements between the warring factions prevented the battle group from deploying until February 1993 when it was allowed to transit Bosnia-Herzegovina and assume control of the Visoko area of operations, northwest of Sarajevo. Additional authority delegated to this battle group included: monitoring the 20 kilometre weapons exclusion zone around Sarajevo and verifying the warring factions complied with cease-fire agreements. Responsibility for sustaining both battle groups deployed throughout the former Yugoslavia rested upon a Canadian logistics battalion (CANLOGBATT) based at Primošten on the coast of Croatia.

Rotations of the individual units comprising the battle groups occurred at six-month intervals. The following CF units served in UNPROFOR from 1992 to 1995:

- 1st, 2nd and 3rd Battalions, R22eR

- 1st, 2nd and 3rd Battalions, RCR

- 1st, 2nd and 3rd Battalions, Princess Patricia's Canadian Light Infantry (PPCLI)

- Lord Strathcona's Horse (Royal Canadians) Regiment (LdSH(RC))

- Royal Canadian Dragoons (RCD)

- 12th Regiment blindé du Canada (12 RBC)

- 1st Canadian Engineer Regiment (1 CER)

- 2nd Canadian Engineer Regiment (2 CER)

- 4th Canadian Engineer Support Regiment (4 CER)

- 5th Service Battalion

Maritime Interdiction Efforts

On the high seas, CF personnel took an active role in the enforcement of all UN Security Resolutions pertaining to the former Yugoslavia. Canadian maritime assets were actively engaged in the monitoring of the UN arms embargo in the Adriatic Sea as early as September 1992. In June 1993, Canadian warships serving with the North Atlantic Treaty Organization (NATO) Standing Naval Force Atlantic (STANAVFORLANT) began enforcement patrols in the Adriatic under the combined NATO/Western European Union operation designated *Sharp Guard*. It's mission was to enforce all UN Security Council Resolutions with regard to trade and shipping sanctions against the Federal Republic of Yugoslavia, as well as the arms embargo against all former Yugoslav Republics. The operation was suspended on June 19, 1996 and terminated following a UN Security Council resolution adopted October 1, 1996. The following Her Majesty's Canadian Ships (HMCS) participated in Operation *Sharp Guard*:

- HMCS *Algonquin*

- HMCS *Fredericton*

- HMCS *Gatineau*

- HMCS *Halifax*

- HMCS *Iroquois*

- HMCS *Montréal*

- HMCS *Preserver*

- HMCS *Toronto*

Air Interdiction Efforts

In the air, from 1992 to 1995 as part of NATO's Operation *Deny Flight*, Canadian air force personnel served aboard the NATO Airborne Warning and Control aircraft charged with the monitoring and enforcement of the No-Fly Zone over Bosnia-Herzegovina. This contribution was revised in September 1993 when two Canadian maritime patrol aircraft deployed to Signonella, Italy to support Operation *Sharp Guard*. In May 1994, after having flown a total of 2,685 hours, Canada ended its air commitment to Operation *Sharp Guard*. Canada was also one of the five nations providing the airlift capacity for the majority of humanitarian supplies flowing into Sarajevo International Airport. Designated Operation *Airbridge*, from 1992 to 1995, Canadian CC-130 Hercules transport aircraft (supported by a 48 person CF detachment based in Ancona, Italy) transported 11,600 passengers and 23,150 tonnes of relief supplies to Sarajevo.

Implementation Force (IFOR)

On December 15, 1995, the UN Security Council acting under Chapter VII of the Charter of the UN, adopted Resolution 1031, which authorized the member states to establish a multinational peace Implementation Force of 60,000 heavily armed troops, under the control of NATO, to ensure compliance with the military aspect of the Dayton Peace Accords (DPA) formally signed in Paris on December 14, 1995.

Earlier, on December 6, 1995, the Canadian government announced that it would contribute a well-equipped force of 1,000 personnel to the NATO-led IFOR. The Canadian designation for this contribution was Operational *Alliance*. Canada's primary contribution to IFOR was a brigade headquarters which provided command and control for a multinational brigade consisting of units from the Czech Republic and the United Kingdom (UK). The Canadian-led brigade was known as 5th Canadian Multinational Brigade. Along with a brigade from the UK, it is one of the principal formations in the UK-led multinational division.

The Canadian headquarters was located in the Bihac pocket at Coralici in Northwest Bosnia-Herzegovina. The combat elements were collocated primarily in Zgon, and the National Support Element (NSE) was located at Velika Kladusa, near the Croatian border. Of the 1,000 personnel, 40 staff officers were employed with the various IFOR headquarters.

As part of IFOR, CF personnel helped ensure compliance with the military details of the DPA by all parties. This included the withdrawal and redeployment of the former warring factions' forces within agreed time-frames and the establishment of Zones of Separation (ZOS). The main elements taking part in IFOR were drawn from Canadian Forces Bases (CFB) Valcartier, Quebec. Units that contributed a total of 1000 soldiers were:

- a headquarters and signals squadron of 264 personnel drawn from the 5th Canadian Mechanized Brigade Group (5 CMBG);

- an armoured reconnaissance squadron of 127 personnel (three Cougar heavy armoured combat vehicle-equipped troops) based on A Squadron, 12 RBC from;

- an infantry company group of 175 personnel (three Grizzly-equipped rifle platoons) and a combat support platoon (mortar, antitank and pioneer section) from the 1st Battalion, R22eR;

- an engineer squadron of 106 personnel (including field troops, a construction troop, a heavy equipment section and a water purification unit) from the 5th Régiment de Génie du Combat [5th Combat Engineer Regiment];

- a small military police platoon of 12 personnel from the 5th Military Police Platoon;

- a NSE of 242 personnel (including an administration platoon, supply and transport platoon and a maintenance platoon) drawn from the 5th Service Battalion; and

- an advance surgical centre of 34 personnel (unit medical station and surgical facility) based on elements of the 5th Field Ambulance and the CF medical group in Ottawa.

Stabilization Force (SFOR)

On December 12, 1996, the UN Security Council adopted Resolution 1088, which continued the evolution of IFOR. Known as SFOR, its mission is to concentrate on stabilizing the current secure environment in which local authorities and international agencies work. SFOR can take military action, including the use of necessary force, to ensure compliance with the DPA.

On December 13, 1996, the Canadian government announced that it would continue to contribute a well-equipped battle group of 1,200 personnel to the NATO-led SFOR. The Canadian designation for this contribution is Operation *Palladium*. Seconded CF personnel also serve with the UN Mine Action Centre in Bosnia-Herzegovina, the UN Preventative Deployment in the Former Yugoslav Republic of Macedonia (FYROM) and the UN Military Observers in Prevlaka, Croatia. Individual staff officers are at time integrated into various headquarters of the UN Mission in Bosnia-Herzegovina. Canada has subsequently stated that it would maintain its troop commitment to NATO-led operation for the foreseeable future.

SFOR Air Interdiction Efforts

The NATO air component providing direct support to the commander of SFOR is known as Operation *Deliberate Guard*. Comprised of a multinational air force numbering approximately 144 aircraft from 12 NATO nations, *Deliberate Guard's* mission includes: conducting day/night patrols over Bosnia-Herzegovina to detect violations of the No-Fly Zone; carrying out air strikes, if required; deterring and/or halting a resumption of hostilities; contributing to a secure environment; and ensuring force protection and freedom of movement.

Canada's original contribution to Operation *Deliberate Guard* was assuming command from Norway of lead-nation responsibilities for the Multinational Air Movements Detachment in Rimini, Italy. This component is called Operation *Bison*, which controls tactical intra-theatre airlift including medical evacuations, movement of troops and conveyance of supplies throughout the SFOR area of responsibility. Operation *Bison* consists of 13 personnel from the 8th Wing CFB Trenton, Ontario.

In May 1996, NATO requested Canada provide tactical fighter aircraft to Operation *Deliberate Guard*. Canada declined at the time, being unable to provide aircraft capable of delivering precision-guided munitions (PGM) for an air support role. The PGM system minimizes the risk to non-combatants. Canada subsequently upgraded its CF-18 fleet, to a limited PGM capability. In August 1997, Canada deployed to Aviano, Italy, six CF-18s fitted with a limited PGM capability and crews.

In order to perform their designated missions, each CF-18 aircraft was armed with live ordnance and was capable of attacking ground or air targets and/or intercepting air targets. This was called Operation *Mirador*. Operation *Mirador* consisted of:

- 110 personnel from all ranks, primarily from the 416th Tactical Fighter Squadron and the 4th Wing, both located in Cold Lake, Alberta. There were 10 CF-18 pilots, seven other officers and 93 non-commissioned members.

Operation *Mirador* terminated in December 1997 when the six CF-18s from the 4th Wing returned to Cold Lake, Alberta. The return of the pilots, and 111 other CF personnel from the deployment in Aviano, Italy, ended a tour in which the CF air component flew 261 sorties and 77 training flights — all without incident — to provide air cover and security of movement for NATO's ground forces, including the 1,200 CF personnel serving with SFOR. As of March 1998, Operation *Bison* continues to carry out its assigned duties over Yugoslavia

Key Events involving the CF in the former Yugoslavia:

- In July 1992, Major-General Lewis MacKenzie was appointed UNPROFOR Sector Sarajevo commander.

- In October 1992, Major-General Robert Gaudreau was appointed Deputy Force Commander, UNPROFOR.

- In December 1992, Canada provided a CF logistics officer to the Office of the UN High Commissioner for Refugees (UNHCR) to assist in the coordination of humanitarian aid deliveries.

- Canada was the first country to respond to a UN request for a preventative deployment of personnel to the FYROM. Elements of the 2nd Battalion, RCR (totaling 175 men) deployed to the FYROM from mid-January 1993 to late February 1993.

- As part of the UN Commission of Experts to report on the evidence of grave breeches of the Geneva Conventions and other violations of international humanitarian law committed in the territory of the former Yugoslavia, Canada provided from 1992 to 1994, a legal officer and a War Crimes Investigation Team which conducted several investigations.

- CF personnel were also instrumental in the first demonstration of the UN safe area concept. In March 1993, the commander of UNPROFOR Major-General Philippe Morillon, asked the Canadian battle group in Bosnia-Herzegovina to assure temporarily the territorial integrity of the enclave in and around the city of Srebrenica, northeast of Sarajevo. At that time, the UNHCR estimated the population around the enclave totalled 44,000; 25,000 of whom were refugees. Within days of the request, 300 soldiers from the

2nd Battalion, RCR deployed to the area to bring the situation under control. The Canadians had a threefold mission in Srebrenica: escorting convoys; protecting the civilian population; and confiscating any weapons within the confines of the enclave. Accountability for the territorial integrity of the Srebrenica enclave rested in Canadian hands until March 1994 when the area ofresponsibility passed to the Dutch Battalion assigned to UNPROFOR.

- In July 1993, Major-General John Arch. MacInnis replaced Major-General Gaudreau as Deputy Force Commander, UNPROFOR.

- In June 1994, Major-General Raymond Crabbe replaced Major-General MacInnis as Deputy Force Commander, UNPROFOR.

- Twelve Canadians died in the line of duty while serving in the field with UNPROFOR and IFOR:

> Sergeant C.M. Ralph, 22nd Field Squadron, August 17, 1992
>
> Master-Corporal J.W. Ternapolski, RCR, March 25, 1993
>
> Corporal D. Gunther, 2nd Battalion, R22eR, June 18, 1993
>
> Corporal J.M.H. Behcard, PPCLI, August 6, 1993
>
> Sergeant J.D.A. Gareau, CFB Valcartier, August 17, 1993
>
> Captain J.P. Decoste, 2nd Battalion, PPCLI, September 18, 1993
>
> Master-Corporal S. Langevin, 12RBC, November 28, 1993
>
> Corporal D. Gavin, 12RBC, November 28, 1993
>
> Private K.D. Cooper, 1st Battalion, PPCLI, June 6, 1994
>
> Corporal M.R. Isfeld, 1st CER, June 21, 1994
>
> Corporal J.F.Y. Rousseau, 12RBC, August 25, 1995
>
> Sergeant C. Holopina, 2nd Canadian Engineer Regiment (2nd CER — IFOR), July 4, 1996

This section is based on public information documents produced by the Canadian Department of National Defence Office of Public Affairs.

Letters from
The Former Yugoslavia

The following letters detail the correspondence of 35 men and women (of varying ranks and occupations) who served in the various missions previously mentioned and wrote to Jane Snailham from April 1992 to October 1997. Wherever the location of the peacekeeper writing the letter is unknown it is noted as "somewhere in the Bosnia-Herzegovina or Croatia...."

The following seven letters were written by a soldier who served with the Canadian Contingent assigned to UNPROFOR:

Sirac, Croatia
April 29, 1992

Dear Jane,

First of all, thank you very much for sending the card and letter of support, you have no idea what it means to a soldier whose mind is being tested every moment. I guess it is only polite to tell you who I am and a little of my background. I was born in a small town in Nova Scotia and finished high school with no future to look forward to. I spent six years in the Militia [in the infantry] then joined the regular force in 1985. My first posting was Winnipeg for three years, then off to Germany in 1988 for a five-year tour. I have obtained the rank of master-corporal and have been on many courses dealing with the infantry. I have seen every province in Canada and been in many countries around the world. And I can assure you that Canada comes out on top. I am married to a wife who goes through everything I do and supports me in whatever I do. I also have been blessed with two boys who were born in Germany [and] who I would die for at the blink of an eye.

As I write, the sun is going down in Croatia and I have no idea of what tomorrow brings. I have never in my life seen or witnessed the death and destruction as I have in this past month. Our company is positioned in a small town [called] Sirac, which is 10 kilometres southeast from Daruvar, which is 20 kilometres south from Zagreb. If you have a good enough map you can probably locate it. The fighting continues seven kilometres from our position. Every night the silence is broken with gunfire and artillery which lights up the sky. The people here are very grateful for the UN being here, but are sometimes confused with our purpose. Everyone from the age of 16 and up carries a gun and whatever else he or she likes to carry.

Our first night here, 232 soldiers from my company were welcomed with 14 rounds of artillery from Serbia into our position. I still have no idea how or why no one was killed. It was the first time since Korea that Canadian soldiers were shelled.

My job consists of policing the area. I have an armoured personnel carrier with one driver and radioman. Every day we go out and patrol the front lines to observe what is going on. At the present time all the UN soldiers in our area have not arrived, only the Canadians, until the remaining soldiers arrive we cannot start working towards peace. I have no idea what the news programmes are saying as I have not heard or saw any news since I left Germany. But I believe in about four months, this is only my opinion, that [the] fighting will stop in Croatia. For the rest of the remaining country, your guess is as good as mine. The men I work with are pretty well much like my brothers. The morale is very high among the soldiers and we enjoy and believe in what we are doing. My stay will only be six months but long enough to miss out on my children's progress.

We are being looked after very well, being fed and equipped with the proper weapons and protective clothing. I find it very difficult to write as I hope you understand. The land here is much like Nova Scotia with mountains only slightly higher. Every town has been destroyed and burnt. Dead animals and the remaining livestock roam the streets; the old money is blowing in the wind. It is really hard to enter a house to check for life when everything is destroyed, usually the family picture is hanging on the wall and it's hard to understand how something like this can happen. Please be very thankful that you live in a safe place. I know [Canada] is being harassed by the language issue and unemployment, but this does not even compare to a lot of places here.

When I finish my tour in October or November 1992, I will return to Germany to my family and pack our bags, then move back to Canada in February 1993. I am planning a trip home to Cape Breton and Halifax in December. If I have time I will stop and say hello.

I must sign off. Tomorrow is a new day, and if God is with us, he will guide us there.

Again thanks for the letter and card, you don't know how it picked up my day. I remember the day after we were shelled, we received Easter greetings from the children in the Canadian School in Germany. It made us all forget the bad and set us in a new direction. Thanks again Jane, and may God bless you.

PS — Keep up the prayers. They seem to be working!

Pakrac, Croatia
May 15, 1992

Dear Jane,

Just a note to say I received your second card and letter, thanks for the reminder. I really don't know what I have said in the last letter so if I repeat myself, I'm sorry. I've been thinking of many other things. Things here are starting to calm down in our sector, but calming down would be saying it lightly!

I have been on the JNA [the Serbian-dominated Yugoslav Army] side and saw the guns that fired on us and in our area nightly. It's hard to believe that within a 10-minute drive, you're on the other side. Their soldiers are better equipped and have a lot more military power.

My job now is to keep open a border checkpoint between the two forces, which is only 100 metres [wide]. In the daytime, only small arms fire is heard; however, at night, both sides come out to play.

My checkpoint is in Pacrac and the city is completely destroyed. In the morning, it's hard to believe children will come out to play in the street, but [they] take off as soon as the fighting starts. I often wonder what's going through their minds. I don't understand why their parents don't remove them from the front line.

The Argentinan, Nepal[ese] and Jordanian UN peacekeepers are starting to arrive in the sector, however, they are not up to strength yet and do not have the proper equipment to protect themselves. I can honestly say the Canadian soldiers here are the best professional army on the ground here in Croatia by far, and I'm not saying that because I'm Canadian. I have spoken to both sides in this conflict, and we have the respect of both to a high degree. You can be sure the Canadian flag flies proudly in Croatia and Serbia. I must go and try and get some sleep, and if you don't get a note from me in a long time, it is because I'm busy.

Anyway, thanks again for the card and the letter, and excuse the handwriting as I'm very tired. The Maclean's magazine people were here last month. Has there been anything in it? I really don't know what the press is saying or if they are even saying anything. Let me know if you can. Take care.

Pakrac, Croatia
June 11, 1992

Dear Jane,

Thanks for the picture and the tartan. I have the tartan beside my mounted gun alongside [the picture] of my two boys. The past month has been quite different.

By the time you get this letter, I'll be in Sarajevo. We got the word yesterday, that we are on a 24-hour [alert] to move south [to Bosnia-Herzegovina] and try and open the airport. It looks like all Canadians will leave this area to take up new residence.

I was in [the village of] Okucani for eight days manning a checkpoint between the Serbs and the Croatians. My partner received a gunshot to the side of the head and lived to walk away which was different. He's okay, but he's not. I won't mention his name as it doesn't matter.

We finally get to see the Nepal[ese] UN soldiers show up, and take over our sector. It's good to see more troops here. I just returned from Austria, spent three days with my wife and two boys who drove from Germany to meet us. My partner came along and met his wife and son as well. I guess we needed a rest. It was different to enjoy three days, but it was the best three days in a long time. All our leave has been canceled until further notice, until we get sorted out in Sarajevo, which will probably take two months, so your picture and tartan you sent me will see the country.

I believe you asked about hobbies, well after getting married and having two boys my main hobby is my family. When I'm not away, I'm with them. However, I do enjoy playing hockey, archery, swimming [and] water-skiing.

I don't know when I'll have time to write again. It depends on how quick we depart here. But be sure, when I have time I'll write. I have a mini-cassette player. If you are interested, I can send a tape with a letter on it. But you will have to get a mini-cassette player to hear it. I use it to send letters to my wife.

Anyway, I must go and get ready for tomorrow. Thanks for the letters. Sorry for the handwriting, I am tired. Take care.

Sarajevo, Bosinia-Herzegovina

July 16,1992

Hi Jane,

A note to acknowledge that I have been receiving your letters of support and [best] wishes. Thanks again for doing so. It sure is nice to receive mail. I get quite a bit from yourself and my family there. I'm sorry you haven't received any notes from myself in the past two to three weeks. I really couldn't come to write to anyone other than my wife.

Our trip to Sarajevo from Croatia was certainly a taste of things to come. We had so much trouble crossing [the border], our main problems came from the Serbs. In a small town called Novo Selo, we deployed to fight our way in or out, whichever came first, but it ended without a shot being fired. The convoy stretched five kilometres, and I was riding up front negotiating crossings [at the opposing checkpoints]. We got here a day late, but all arrived safe and sound.

There is really nothing good that I can say, other than the mail I get or a cold beer before I go to sleep. My job is to go out at 0600 and occupy an intersection and send reports [to my] higher [headquarters]. This is to ensure that there is no fighting at that point, so the convoy of supplies can cross safely, to say the least. I have been shot at so many times that I'm wondering if I'm invisible to man, not only myself, but the rest of us that man the crossing points.

The sight of dead animals in Croatia has been changed only to the sight of dead humans who litter the gutters and river which flows through the main city. I don't know if people just forget them or what. Normally, you would think there is no war here, if you were deaf, as people walk along the streets, only to break into a run between the buildings in fear of being shot. It's unbelievable to sit and watch people (regardless of their sex or age) be gunned down in the street. The sniper's favorite target is the buses, as they are full of people and you would have no trouble hitting someone. It is hard to look into the eyes of an old man, and try and tell him he's okay while you try and patch him up the best you can. We transport them to the UNPROFOR hospital for quicker response. This is only one situation of many that happen on the hour!

Two children were killed while talking to the soldiers in front of our headquarters. Children [usually] gather [there] to receive candy and talk. Last night, the Serbs launched a mortar attack on the main street. The bombs land[ed] where the children were standing. Eight were injured and two dead.

At the airport, it is no better. The Serbs shoot at airplanes loading and taking off. We had no other choice but to deploy our own snipers to protect the relief flights from further attacks.

With all this happening, the relief supplies are still flowing. I often wonder if it is worth it. As of this date, no one was killed yet, but we have had some serious injuries and gunshot wounds. We all know someone is going to get it, it's just who and when, but it's going to happen. When I first got here and experienced the baptism by fire at myself and my friends, I was shocked, but now we no longer worry about getting hit. Its kind of like going to a circus when you were a child, and you were scared of a certain ride at the fair. Once you got on the ride, you lose the fear and end up expecting it.

The Egypt[ian] and Ukrainian soldiers are starting to arrive as well as another French battalion from France. As soon as they get on the ground, we will be departing to return to Croatia. This may happen before the end of the month. When I get back I'll get two weeks leave, and I'll head back to Germany to see my wife and kids. When I get there, I'll give you a phone call.

The time is passing by very fast. We work 14 hour days, then head back to our sleeping quarters for [our] orders for the next day, then try to relax and go to sleep.

Our camp is hit nightly with small arms fire and the odd artillery or mortar round. We have lost a lot of vehicles so far, but nothing serious.

I am really sorry as this is not a letter for anyone to read, but it helps to get it off my chest, as we don't talk about it among the soldiers. I wouldn't dare say this to my wife, so you are helping out a lot by reading this whether you realize it or not.

I can't send any pictures as we have no film for the instant camera left. I'm talking lots of pictures, but there is no place for development. Regardless, I take slides, so when I get back to Canada I'll drop by and show them to you. I'm sorry if you asked questions, it's hard to remember.

Well jane, I must stop babbling and get ready for tomorrow. Sorry for a depressing letter, but if I didn't say what I said, I would only be lying and making up shit. Anyway, I'll sign off and I'll write again when I have time. Thanks so much for the support and prayers. May God bless you.

------------------------------ ◆ ------------------------------

Somewhere in Croatia
August 6, 1992

Dear Jane,

As I write this letter , I am sitting on an observation post (OP) in Croatia, leaving behind Sarajevo which will never be forgotten. I am sorry I have not written more in the last month. I had no words to write about and even found it hard to write my wife. That [is] all in the past now and we are getting on with a new mission here in Croatia. It was really different to roll back into this area, as all the heavy guns have been taken away on both sides along with the soldiers. There are no more of either within the UN Protected Area (UNPA). It is hard to stand the quiet, it seems to be a new part of the world. Our main role now is to patrol the woods by night and day to ensure there are no insertions by either force. It is still very tense here between the Croats and Serbs as the resettling there begins.

By the time this letter gets to you, I should be in Germany on my UN leave in August. The latest letter I received from you was the one with the news clipping. I think it was pretty neat that they came to the house with the TV camera. My hat goes off to you for the support and the letters you have sent in the past four months. You mentioned that we would not get danger pay, It didn't surprise me or the rest of us. Just being alive at the end of the day is enough reward for us. Besides, your letters are worth more to me than danger pay.

The weather is now very hot in the day, we try and drink as much water as we can. So far, we haven't had any problems with the heat. We were lucky to arrive in April and gradually be exposed as it got warmer. I'll be sure to get a picture off to you as soon as I get my hands on one. I'll be developing some pictures and I'll be sending you some.

The events in Sarajevo are beyond words. I had believed that I had seen it all in Croatia. I just pray to God our children and people never have to witness the death and destruction caused by war. Being shot at by everything from rifles, rockets and artillery, I got used to, but look[ing] into the eyes of starving, frightened children who have absolutely nothing to do with this war turned my world upside down.

Things have happened to us that I cannot explain. Why we did not come out with casualties? A lot of [my] group call it luck, but I'm convinced God was looking after the Canadians in Sarajevo. Well Jane, enough of that, I'll sign off and I'll be sending pictures in the next letter. May God bless you for your efforts in helping me keep my strength to carry on.   I hope the weather in Halifax is nice and warm and you keep in good health. Thanks again for your support. From your true friend.

**Somewhere in Croatia,
August 26, 1992**

Dear Jane,

Greetings from Croatia. I arrived here last night from Germany after two wonderful weeks of time off. It was sure nice to be in civil land again. I got five of your letters which were laying on my air mattress when I returned. I'm sending [you] a monthly news book which is printed here for ourselves. Since the French are in command of the Battle Group, there really isn't too much English in it, that's the French-Canadians for you. Also, there is a photo taken while returning from Sarajevo. I'm driving the vehicle so you can only see part of my head. At the time of the photo, my driver was getting some sleep, as the road move was 40 hours long, but it is my carrier.

As I write this letter, I am on duty in the command post. This is where we control all the patrols who are out for the day. They radio back to let us know their progress, etc. Usually, we work a 12 hour shift, then patrol for five days. It breaks up the time. There is no shooting as regular as Sarajevo was, but we have a problem at night with people coming back to live in their homes, with people threatening them if they stay.

All the military has left our UNPA which is an area controlled by the UN. We set up roadblocks and search cars for automatic weapons which are forbidden in this area. If we find [weapons] in the vehicles, [we] take them, [although, the occupants are not] very impressed about it. Because the UNPA is now open for people to come back, there are many reports of harassment by both sides.

I believe I wrote to you to say I'll be going to Nova Scotia on my next posting. There is word now that the battalion there will be coming back here next April. I don't know how my wife will take that. It hasn't been confirmed yet, but it sure looks like it. Now that I'm back, I can finally unpack my equipment which has been packed since I got here. We were never really in one spot long enough to unpack.

Well, I must sign off now, so I'll say good bye for now and take care.

**Somewhere in Croatia
September 3, 1992**

Dear Jane,

Things are very busy. We are having problems with people in possession of weapons. Trying to take them away is difficult.

The weather is starting to get cool, especially in the night. Usually, I just sleep on top of my sleeping bag. But last night, I woke up fighting to find the hole. By the time this letter gets to you, I'll have probably only three to four weeks left in-country. Time sure flies fast, when you are having fun.

The soldiers from the PPCLI who are replacing us, their advance party people have arrived to look at the place. We are pretty lucky as for staying in a building; there are still a lot of soldiers sleeping in tents. The winter is coming, and people better start working fast to help them out as for accommodations.

We also have problems with contractors not wanting to work for us, as [the] UN has not paid any bill yet for work done by locals, which is not making things easier. They have sent down from Germany some lockers and furniture which was going to be thrown away from the base closing, so that helps out a bit. Our medals parade is on the September 15th. So that will be a blast. Most of us don't want to go through the bother of getting dressed up to partake. I would rather receive it coming back from a patrol and find it on my pillow. But regardless, mine will be going to my wife. Anyway, I must go and I'll write again soon. Take care.

**Somewhere in Croatia
September 11, 1992**

Dear Jane,

Hello. I can't remember when I wrote last. However, by the time you receive this letter, there is no need to write to me here in Croatia, as by the time I receive the letter I'll be back in Germany. However, I have no problems with continuing writing to you.

It looks like I'll be heading to a base in Nova Scotia in February 1993. The regiment I will be joining is getting ready to deploy to the Split-Sarajevo sector. I've been told that I may just end up coming back over in March. The books say that after a six month tour you cannot go back on another tour for one year. But the book also says if I am needed I go. So I guess I will have to wait and see what happens. I really don't want to think about it now.

We are expecting problems here before December, but we really don't know when. This war is far from over.

You probably heard two French soldiers were killed in an ambush. I know where that ambush spot was, just directly outside the airport. We were also hit there many times. It is just too difficult to locate where the fire is coming from. You can shoot back in the general direction with lots of fire-power, however, we were always worried about killing women or children who lived in the area. Maybe we were crazy for not doing so, but none of us died. [So] there is no reason to say that we did wrong.

It was the Muslims who killed the two soldiers and it was also them who shot down the Italian plane [bringing in relief supplies to Sarajevo airport in early September 1992]. Now you can understand why General MacKenzie tells it like it is. This is what we get for trying to help these people, and why I hate them all. I don't know if I ever told you, but the Serbs have never shot at us with artillery or mortars in Sarajevo. It was the Croatians or Muslims who pounded us daily. Why? To get the people of the world thinking and hating the Serbs more than blame it on them. I wouldn't say that if I didn't have proof. But I do.   The replacements will start arriving within the week. I haven't got a date when I'm leaving, but I'll be back in Germany before September 6th.

If you write me in Germany, I will certainly write back. You have done a wonderful job in helping me keep going. I can't thank you enough for the support. You don't know how much it means. Thanks again. I'll write another letter to you before I depart.

◆　◆　◆

The following are two of three letters written by a soldier who served with the Administration Company of the 3rd Battalion, PPCLI assigned to the Canadian Contingent of UNPROFOR:

Somewhere in Croatia
November 25, 1992

Dear Mr. and Mrs. Snailham,

I just received your heart-warming letter, and I must say that Canada is the best country in the world, and its people like your family that makes Canada the way it is. I was a bit confused as to who you were because I tried to think who I know or should I say where do I know you from, but when I opened your letter, I realized your family is part of the special group of concerned Canadians who think about the soldiers away from their homeland and families. It's not too often that someone outside of the military family takes time to write or think about what the CF are doing and where they are. You are right, people don't know what the CF are doing, and the statements like, "its their job or where are they now" just confirms my point that there are people that just don't care. The Canadian Maple Leaf has taken a beating this year from these people, but we will make it.

I am from a military family. My father was in the PPCLI as am I, so the tradition carries on. I have lived from coast to coast, but most of my years have been spent in Nova Scotia. I now live in British Columbia. My wife is an accountant and we have been married for just over two years. We have two cats, one that is very overweight, but we just say she is holding water when people ask us about her weight. We have no children, but the neighbors make up for that. I retire from the army in September 1993, because I have reached my age limit, which is fine for me because after 20 years doing the same job, it is time for a change. But my heart will always be [the] army.

I missed the Blue Jays winning, but all right, go "Jays." The hockey games are now coming in, but they are around two weeks old and the joke over here is "who wants to make a bet on the game," as we already know who has won two weeks ago from the news. Things that you see on TV about Yugoslavia, about the killing and destruction, are true, but they don't show it all. Our Canadian soldiers everyday are in danger one way or another by the Serbs or the Croatians. I will say that it is not a very kind or happy place to be. We could all be going home in March 1993 as the Croatian government has so far refused to sign the mandate which has to be signed by March 3, 1993. They want us to go home and leave them alone. I say, so be it.

I don't mind if you write and send cards. I will respond also if you and your husband don't mind.   And you never know, we may get a chance to meet as I and my wife have talked about going to Nova Scotia for a visit. I took her there in January 1990 when my mother died, but you know how [the] winter can be, we never stayed long. But she has wanted to go back to "God's Country." And again, I thank you for your kindness and thoughts and prayers. And bless you from my family. The "3rd Battalion, PPCLI" family that is.

***

Camp Polom, Croatia
December 29, 1992

Dear Jane,

Hello from a soggy Canadian soldier. Yes, it is raining over here too. We had some snow, well, quite a bit of snow. Everything was white for about three days, but it rained also and it was pretty cold on top of that, so we just grin and carried on with our work. Things have been quiet. I guess with Christmas coming everyone will be taking time off [from] fighting which will be nice. They really don't have Christmas like we do. The stores are bare, not too many toys for the kids or decorations. We brought most of the decorations from Canada before we left. So we are all ready. I try not to think about my wife. I know she will be alright as she will be spending the Christmas season with her family in Ontario. It will be a quiet time for myself, but it is only one day and it will be over.

The people over here assumed the wrong idea about what our jobs were. The people think we should be involved with the war, but we are here to keep the peace. They are very mad at us for this reason, and they want us to go home. There is an election in January 1993 and the Prime Minister that is in power now [Franjo Tudjman] said if he gets back in, the UN will be sent back in February. I have talked to a few people about this and the people agree with him. So, I really don't know what to say about it.

I'm outside of Daruvar, approximately 10 kilometres [from it is] Camp Polom. It is a camp that used to belong to the Serb army, and when they left they took everything, including the heat. But we are Canadians and we will overcome. Well, I will close for now and write again soon.

PS — Merry Christmas and a Happy New Year to your family.

◆   ◆   ◆

The following three letters were written by a 27-year-old sergeant who served with Golf Company, 2nd Battalion, RCR assigned to the Canadian Contingent of UNPROFOR:

Somewhere in Croatia
January 26, 1993

Dear Jane,

Thank you very much for writing and keeping me in touch with what's going on back in Canada.

I'll tell you a little about myself. I was born in a small French Acadian town in Cape Breton. I'm 27 years old and I've been in the service for seven-and-a-half years. I'm married and my wife is also from Cape Breton. We have a son who turned three years old on December 11th. Both my wife and son are in Cape Breton while I'm in Yugoslavia. We have known each other for over six years and have become really good friends. Me and her and the boy will be going to Cape Breton shortly after they repatriate back to Canada from Germany.

The military has taken my family and I to Winnipeg for two-and-a-half years, then to Germany for four and now we are posted to a base in Nova Scotia. Over here in Yugoslavia, our daily routines are as follows: two days of camp security, two days on OPs and two days of general duties. We always have troops on leave. Either 60 hour passes or UN leave. [During] my leave, I will be returning to Canada to be with my family.

I returned last night from an OP we have set up two kilometres south of our camp. It's there as an early warning for the camp, and we also call in any mortar or small arms fire we see. I've received more mail from you in the last few days and the Nova Scotian tartan colors are nice. Thank you.

I'm sorry its taken too long to reply.   I'm probably going to call when I get to Halifax when I'm on [my] UN leave. I don't have much more to say for now. Take care. Thanks for writing.

Visoko, Bosnia-Herzegovina
March 6, 1993

Dear Jane,

Hello and how are you doing? The weather over here is very cold and snowy. We are now in a town called Visoko, about 18 kilometres northwest of Sarajevo. We are doing our first large convoy escort on Monday and it is expected to take five days in order for us to get through and deliver the supplies.   The guys are going into Sarajevo all the time, escorting small military convoys or for recce's (reconnaissance) and it is bad as everyone says.

I just arrived here today to find two letters you sent on my bed, so it was a nice thank you. Everyone in this town carries a weapon and people all over want to sell us things, "anything." Hopefully now that we're here, the taxpayers' dollars will go to good use. Then again, not too many people would put up with the "shit" we do anyway. Since we got to Visoko, people have opened their eyes and are realizing there are people who are starving and it's easy to pick them out of the crowd. Well, good night for now. Take care and I'll write again soon.

PS — I'm sorry if I'm jumping from one topic to another. I'll try to make my next letter a lot clearer.

Visoko, Bosnia-Herzegovina
April 2, 1993

Dear Jane,

Thank you very much for the card and letter. To answer some of your questions.

1. We heard about the celebration that's being planned for us, however, I just want to be with my family and friends, and not worry about any parades.

2. I think on return to Canada, we will land in Saint John and go through customs there and then bus to Oromocto, New Brunswick (NB) where we'll do our final clearing and then go home. I don't think that there's any chance of landing in Halifax because we flew out of Saint John to come here.

3. One convoy I was on made it on *CNN*, but it only showed one of my soldiers, not me. We were stopped at a Serbian checkpoint for four days and we didn't get through. Our wheeled vehicles were re-routed and we returned back here.

I am going into Sarajevo tomorrow for the day, and our medals parade is on Monday. On Tuesday, we are escorting a convoy southeast of Sarajevo to a town called Gorazde. Hopefully, we [will] make it through. I will give you a call when I return, and I do make it to Halifax every once in a while because I have a brother in the navy.

Once again thanks for the letters. Well, that's it for now. Take care.

◆   ◆   ◆

The following is the first of two letters written by a corporal who served with the Administration Company belonging to the 2nd Battalion, PPCLI assigned to the Canadian Contingent of UNPROFOR:

Camp Polom, Croatia
June 4, 1993

Hello Jane,

All is well here and things are very busy. My partner and I have been inspecting weapons in each company. So there's lots of driving and repairing weapons. By the way, I know I told you I was a weapons technician, but I'll tell you what that means. I repair all land weapons, for example, pistols, many types of machine guns, mortars, shot guns, tanks and [M-]109's [an American-made 155mm calibre self-propelled artillery piece] which is the largest artillery piece and weapon I work on. I also repair heaters, stoves, lanterns and propane burners. There's also other little odds and ends, but I won't get into them. I like working on big pieces of equipment compared to little ones. I enjoy my job, but sometimes...well you know.

I am still in Croatia, but there has been a lot of talk about us moving to Sector North. I thought north meant up, but its southeast of here. I won't believe it until its actually happening. We have a good set up here, and I would hate to set up a new camp. There's too much work involved.

We have line tours once in awhile and they're quite incredible. A line tour is for a person usually from Camp Polom, going to a rifle company and becoming an infantryman for three nights. The person does [a] foot patrol, wheeled patrol and checkpoints. I've already completed one line tour and it was exciting as well as scary. While I was on foot patrol, there were bombs going off, machine gun fire and even a grenade which blew up a car. All this happened in Pakrac with A Company. There are four companies: A, B, C and D [Alpha, Bravo, Charlie and Delta]. I remember when the first mine went off, I was on foot patrol and we all dropped to the ground. Things were pretty scary after that — I wasn't sure what was going to happen next, but at least we were ready. I'll probably do another tour at the end of June. It will probably be exciting just like my last one.

People in Croatia seem to be fairly nice. Most of them wave to you when we drive by. If they talk English, they'll talk to you. All in all, they don't seem to mind us being here.

Well Jane, take care of yourself. I'll write again. Thanks for your support.

◆　◆　◆

The following two letters were written by a 2nd Battalion, R22eR Major assigned to the Canadian Contingent of UNPROFOR:

Visoko, Bosnia-Herzegovina
June 11, 1993

Dear Mr. and Mrs. Snailham,

I am the commanding officer (CO) of "C" Company of the Tactical Group of the 2nd Battalion, R22eR. I [will] take a few minutes to answer your nice letter. Please excuse my writing, since my first language is French.   I have a family of three, [including] my wife, my daughter of two years old and my son of one year old. We all live in a bungalow near Quebec City in a town called Ste. Catherine de la Jacques Cantier.

I am in Bosnia since April 28, 1993 with my company, which is 114 men strong. Most of them are infantrymen and officers. We have spent the last month in the enclave of Srebrenica. It was a great experience for all, and I have learned a lot about this war. It is important to tell people that this war is not for a religious question, but for land, natural resources and basically, money. I had many occasions to meet both parties. I had the chance to talk to [the] Bosnian people more often because it was easier. As [a] Canadian soldier, [I] find it hard to accept someone [who] kills men, women and children who don't bear arms. Thank you for your support.

◆　◆　◆

Visoko, Bosnia-Herzegovina
July 12, 1993

Dear Mr. and Mrs. Snailham,

This war is not about religion, but about land. The Serbs and the Croats saw a chance to enlarge their territories at the expense of people who were not ready for war. In [the] Srebrenica enclave, you find a lot of mines: zinc, iron [and] lead. There is still a complete [plant] manufacturing chair and tables, complete with millions of dollars of equipment. It is still the subject of fighting. In this county is the biggest bauxite mine in Europe. Also the fact that the Srebrenica enclave as [well as in] Zepa and Gorazde, those enclaves in [which the] majority [are] Muslims (60 per cent) are close to Serbia.

We are in county Visoko, where peoples of different ethnicity live and fight together. I have met them on many occasions. There are a lot of officers in the Bosnian army who only want to live in peace in their country with everyone. There are those wildhead extremists [as] they call them here, who believe in a Muslim country, but they are few. But they are getting support from Muslim countries because the rest of the world doesn't care. The people have nowhere to go, so they will fight to the end. I have seen many kids without proper clothing or food. I find it hard because I have two children myself. About the Canadian soldier who was killed, all I can say is that the Serbian was trying to hit the vehicle, but they killed him. My men were at the same place that morning, but God chooses the one he wants. Everyone knows the risks when we signed up. We are trying to do as much as possible with the equipment the [Canadian] government is giving us. It would need a lot more soldiers and equipment to keep the peace in this country, but we (the world) are not ready to pay the price because we don't have economic interests to protect. I should be back in Canada [in early August] because I am [being] posted to the Canadian Airborne Regiment at CFB Petawawa, Ontario with my wife and kids. Thank you so much for your support. That makes me feel great to be a Canadian. Je me souviens. [I remember.]

◆   ◆   ◆

The following letter was written by a 25-year-old captain who served as a pharmacist with the medical detachment assigned to the Canadian Contingent Support Group assigned to UNPROFOR:

Camp Polom, Croatia
July 25, 1993

Dear Jane,

Thank you so much for your letter. Yes, I am female, but I'm not a nurse or medic. I'm a pharmacist. We have no nurses in Croatia but there are four or five of them with the Forward Surgical Team with CANBATT Two in Bosnia and our female medic was away on a convoy when your letter arrived. Since then, she returned to Canada as her rotation was finished. Sorry I haven't answered earlier, but I'm a very efficient letter-writing procrastinator.

I am from British Columbia (BC). I was born in Prince George and grew up in Smithers. I have two younger brothers and two younger sisters. I am 25. After 12 years in Smithers Christian School and Buckley Valley Christian High School, I went to Okanagan College for two years in the university transfer program (for honours in Biochemistry). In the summer, I joined the reserves (BC Dragoons — an armoured regiment) which I really enjoyed and then in 1987, I joined the regular force Reserve Officer Training Programme — where the military paid for my university and I pay them back with four years' service afterwards. I went through the Basic

Officer Training Course in Chilliwack from July to August 1987 and then started in at the University of British Columbia. It worked out that I was able to transfer into the second year of the pharmacy programme, so I didn't lose any time. I graduated in 1990 and got my license in June. I had to take a couple military courses (about medicine and medical resupply in the CF) in [CFB] Borden and Petawawa that summer and then was posted to the CF Hospital in [CFB] Valcartier which is just outside Quebec City. Not being bilingual, it is rather difficult to do my job. (I'm finding it difficult to write with this pen, so hang on while I switch! Ha! Much better!) But that year, we seem to have had a surplus of pharmacists, so I was in a supernumerary position and was less essential. After five months, I got an offer to work in the base hospital in Petawawa, while their pharmacist when to Saudi Arabia during the Gulf War. I decided to go to find out what I really knew since I'd be the only pharmacist working in English. After the Gulf War, they asked me to stay since the other pharmacist was being posted out that summer anyway, and I stayed. That job was the busiest and most challenging I've ever had (so far!). I was part of the 2nd Field Ambulance, which is the medical unit which looks after a brigade, thus, I was the brigade pharmacist for approximately 4,000 troops. Second Field Ambulance was also tasked to man the base hospital, thus, I was also the base pharmacist. It was an excellent experience and included one three week exercise in the Petawawa training area living in the "hootchies" and wearing camouflage paint, helmets and bug nets all the time. It was a learning experience! It also included *Rendezvous 92*, a division-size exercise (totaling 13,000-15,000 troops) in the Wainwright delta for two months in the spring. There, I was the division pharmacist, and was looking after medical resupply to about 30 different units and sections.

In the fall of 1992, I was posted again. This time to CMED (Central Medical Equipment Depot) which looks after the purchase and distribution of medical drugs, dressings, equipment and related items to the entire CF as well as the manufacture of first aid kits, field ambulance treatment sets, medic kits and major air disaster kits. I was put in charge of the manufacturing section. I was quite disappointed to be leaving dispensing so soon, but I figure I got the best job at CMED, so it wasn't too bad.

The biggest payoff came with this UN tour. CMED sent the first pharmacist here and was tasked to send her replacement. Because I'm quite young, I had to convince the commanding officer of CMED that I could do the job and that I really wanted to go. Frankly, he need have no worries, because, overall, I'm finding this to be quite a breeze, as well as exciting, and my varied experience certainly helps. I still wish I could have come on the first tour — when they were building everything from scratch, making all the contacts, and setting everything up. This tour, the set up is all done and all that is left is the fine tuning — and the continuing problems which arise due to the changing situation in Bosnia and Croatia.

I came here on the first of May 1993, and will still [be here] till November 1st. I am in charge of the FMED (Forward Medical Equipment Depot) with four others on staff, one of whom is usually down in Bosnia with the Forward Surgical Team which has most of the medical equipment he has to maintain. FMED is one section in [the]

CCSG (Canadian Contingent Support Group) which is to provide logistical, mostly resupply, transport and maintenance support to CANBATT One and Two — the two infantry battalions from Canada which are in the former Yugoslavia. CANBATT One maintains the peace in a UNPA in Croatia called Sector West. Their headquarters is in the north [half] of Camp Polom and CCSG is mostly in the south camp. Camp Polom is just outside Daruvar, which is about two hours east-southeast of Zagreb. CANBATT Two is the battalion in Bosnia-Herzegovina — in Visoko and Kiseljak (near Sarajevo), and in Srebrenica. CANBATT One requires little support and what they require is easy because they're right here. CANBATT Two is much harder. We can't just drive to them through Bosnia because of the fighting, so we get to Visoko over the mountains from the southern coast of Croatia and to Srebrenica via Serbia. Both these convoys tend to be fraught with difficulties!

Being here sure helps me to be aware of what the situation is here. Back in Canada it just seemed very confusing, and really it is, confusing with people changing sides depending on who's winning and the various alliances throughout Bosnia-Herzegovina: Serbs and Croats, Croats and Muslims or all three against each other, the mercenary Black Swans [a special forces unit of the Bosnian army] who are just there to fight and the real Muslims who came thinking this was a religious war — it's not. The Muslims here are like the majority of Christians in the western world whose Christianity is a background/ancestry not a religion. The war seems to me to be everyone trying whatever means they can to get a piece of the pie: or all of Bosnia-Herzegovina and good portions of the UNPA's in Croatia. All sides are determined and none are truly willing for compromise — no one will back down. One guy in Visoko said to me, "when you go (the UN that is) we will be able to get rid of them and then there will be peace," which is ludicrous because every side says that same thing, and they can't all win, and certainly all are losing.

Well! I seem to have become quite verbose. I hope I haven't bored you, but you asked! If you want to write back again, I'd love to hear from you.

PS — About your concern about my being treated with respect and honour — from Canadians, there is no big need for concern happily; some are treated with less than others, but truthfully, sometimes it is justified. If we are honest, we will admit that just like there are men who do not deserve or earn much respect, likewise, there are such women in existence! From other nations, the most problems are [from] the locals, who don't want the UN here (male or female) because they [the locals] want to fight. So, we have a curfew to prevent problems. We have to be in camp by 2200 — a reasonable precaution.

●　●　●

The following letter was written by a 32-year-old master-corporal who served with the Administration Company of the 2nd Battalion, PPCLI assigned to the Canadian Contingent of UNPROFOR.

Daruvar, Croatia
July 28, 1993

Dear Jane,

Well Jane, I guess I'll tell you a little about myself. I'm originally an Ottawa girl and I, for nothing better to do, joined the military as an administrative clerk at the age of 19 in December 1980. I guess I became tired of the cashier and waitress scene and decided to try something different, besides my mother said I'd look good in a uniform. I'm the oldest (now 32 years old) and only girl in my family amongst five younger brothers, the youngest being 24 years old. In 1983, I married a civilian Newfoundlander (worst mistake I ever made) until my divorce in November 1991. There were no children out of the marriage which I am grateful for.

Since I've been with the military, my postings have been to CFB Borden, Ontario [for] four years (my worst posting), Ottawa — four years, Canadian Embassy, Washington DC — two years and then Winnipeg since July 1991. My best posting was Washington for the following reasons: my husband left me, I met the new love of my life (he worked at the Pentagon) made so many friends, met different people and traveled extensively. He's also with the military, the same trade as myself, from Ottawa and joined the same time as me. It's strange that it had to take a posting to another country to meet the man of my dreams. His parents sent me a birthday card and they signed it off as "Happy birthday to our future daughter in law." I think this is a hint of things to come. Both our families live in Ottawa, so we were talking awhile back, that if we did get married it would be in Ottawa in a church. I already like the idea of having a two week Club Med honeymoon in the Caribbean. Right now, though, we're only common-law and I won't be ready for marriage for at least another two years.

Both he and I were transferred to Winnipeg at the same time we decided to purchase a house together. It's a nice little bungalow with three bedrooms, two full bathrooms and a finished basement. Mind you it does require some renovations: like the kitchen, the main bedroom and I'd like to see new carpets. Because of the type of job I do, it's very difficult for me to start a garden, (I've always wanted one) or to do any type of landscaping.

When I first arrived in Winnipeg I discovered that the city didn't have a Volksmarch club [a 1-kilometre, cross-country hike followed by a mini-festival], so in all my eagerness I decided to start one. I hadn't realized that a large undertaking such as starting Manitoba's only Volksmarch club was going to restrict my social life to a point where I don't have a social life. But I persevered and low and behold, our club is going into its third year.

My [other] hobbies are scuba diving, camping, needlepoint (mostly cross-stitching) reading and traveling. I've done my fair share of traveling since I joined the military, but I never tire of it. I'd like to plan a camping/backpacking trip next year to Greece. They have the most wonderful trails and sites. My love wouldn't be able to join me because he has very bad knees and we plan on doing a minimum of 30 miles a day of backpacking along the trails.

When I return to Canada, I seriously have to consider starting up on some courses and work another career. I'll be retiring in seven years and I'm not ready to go into the real world. I still have no upgraded education or experience except for my typing and computers, but I don't want to do that all my life. I was thinking of getting into forestry and wildlife conservation or become a scuba diver instructor or maybe both, I haven't decided.

Just a little geography. Camp Polom is situated about five kilometres from Daruvar, Croatia. Daruvar is about a two-hour drive southeast from Zagreb, Croatia, but our mail that comes from Canada goes through Belleville, Ontario and is flown over twice a week. Our situation here is deteriorating, that's why most of our battalion has deployed south. Sometimes we have Croatians that do get restless and tend to get a little violent by throwing grenades at the camp or fire their weapons and there have been gang beatings when entering the town. Most of this violence comes from the young Croatians probably with nothing else to do. They no doubt see us UN soldiers as a threat. I just assume leave this country and let the Croatians and Serbs continue with what they want most and get it over with, genocide. After all, it seems that they have their heart set on it. We're just in their way. Sorry for being so negative, but I think you'd understand if you've had to live around these people. These people just don't comprehend that their little war has taken me away from my family and home.

I don't mean to be so pessimistic, but this place is starting to get to me. Not much longer before I head home so I keep reminding myself.

My UN tour here is a little hectic. Twice a week, I have a duty called a roving patrol. At all hours of night, two of us on duty have to patrol Camp Polom's three gates and also the interior of the camp. This certain duty was started due to threats and gun and mortar fire around the camp.

I live in a small trailer with another girl, and our shower and bathroom facilities are across the camp. There are approximately 30 military girls and over 1,000 guys camped here with me. Mind you, the guys are spread out around the camp in different areas. Some of the trades that the girls are in are, medical assistants, pay clerks, administrative clerks (like myself), vehicle mechanic, truck drivers, military police and cooks. There are still a lot of men here that don't believe that women should be allowed in a combat zone or wear a uniform, but if women can do the job just as good as the men, why not. The camp's kitchen and drinking mess are located in tents, like a MASH [mobile army surgical hospital] unit, and my office is in a building riddled with bullet holes. The Croatian army had control of this camp before the Canadian UN troops moved in, so most of the buildings have either been blown up or shot up. Our construction engineers teams are still making repairs around the camp. The odd time our vehicles hit anti-tank mines that have been planted in the road, there were no serious injuries.

The insects here are very bothersome, especially the mosquitoes and ants. We've tried ant traps, but the pesky little vermin seem to enjoy living in them, so I've resorted to spraying them with insect killer. So far, no sign of ants, but we shall have to see...

[On] my second trip, I had gone to Istanbul a few weeks ago. It was a long one, but good... We visited the Blue Mosque (a Muslim church). After our visit there, we walked across the street to visit a sultan's palace that used to house 5,000 people. We saw some of the sultan's treasury that included one of the largest diamonds in the world, everything was made of jewels and gold. Our tour during this day included a rug show (and what beautiful rugs) and a leather fashion show at a department store. The clerks there served us beer, tea, schnapps and Turkish coffee. While shopping at the Grand Bazaar, I came across a really good deal for a crystal mosque for 26 DM ($20.00 CDN). I also purchased, for my little niece, a 14kt gold chain and little gold crucifix for her 1st birthday and for my "huggy" bear, a chain, a crucifix and a puzzle ring but he doesn't get them before Christmas.

...I would have liked to have seen some of the countryside, but the tour guide stated that it was a couple of hours ride outside the city and that it wasn't part of the tour package. To say the least, I was very disappointed.

Last Friday, my roommate and I went to Zagreb, Croatia to spend the night and do some shopping and looking around for a day downtown at the market square. All weekend there was a folk festival with a lot of different dancing and singing from the local Croatian people. The costumes were very nice and colourful and there was even American dancers that were visiting from Seattle, [Washington, USA]. Have you ever been to Seattle? It's a very interesting city and I'd like to visit again someday. Saturday morning we went to the market place to pick up a souvenir and what a crowd. There was standing room only. We did a lot of walking around then checked out of our hotel at 1300. We waited for our ride which was to come and pick us up at 1500, but no one showed up. I called back to camp and was told that everyone has [moved] down south to Bosnia and that we were to return on our own to camp. [My roommate] and I ended up taking a train back to Daruvar, then we hitched a ride on a farm tractor that took us back to camp. We walked into camp and discovered that most of the people had already left. I will also be going to the war zone when I return from UN leave next month, but not before.

So far as I know though, we are still returning to Canada on schedule and the date I fly home is not soon enough for me. I can't wait to sleep in my own bed and eat proper food. I look forward to going to the bathroom under the same roof I sleep under. Ever try to go to the bathroom in the middle of the night in a place like this. I have to get dressed, make sure I have Kleenex with me because most of the time there isn't any in the johnnies (portable toilets) and then make my way to one of the portable potties. It's worse when it's pouring rain out, that means I have to put my combat boots on along with my rain gear. By the time I return to my trailer, I'm wide awake. I've gotten out of the habit of drinking lots of water before I go to bed just so I can avoid such things.

I have a 17-day UN leave coming up next week and it will be the first time I seen my love in four months. He is flying to Vienna, [Austria] where we'll meet and stay for a couple of days. We're taking the Eurorail from Vienna down to Southern Italy and then a ferry from there to Athens, Greece. We are going to spend a week or so on the Cyclade Islands in Greece. I've heard all kinds of great things about these islands, so we've decided to try it out. If I really like it there, I still want to go next year as a backpack expedition.

My love has been great the whole time I've been away. I hear he's been doing some work in the house and keeping himself busy. I always brag about him and how wonderful he is compared to all the neanderthals (men) in this camp. I usually keep busy myself with needlepoint projects, exercises and I also have a lot of reading to catch up.

Thank you again for taking the time and writing to us. I hope this letter will give you some insight on our situation. Take care and I hope to hear from you soon.

◆   ◆   ◆

The following letter is the second of two letters written by the corporal who was mentioned earlier on pages 20-21, and who served with the Administration Company's belonging to the 2nd Battalion, PPCLI assigned to the Canadian Contingent of UNPROFOR:

Kijevo, Croatia
August 3, 1993

Hi Jane,

Myself and approximately 450 other personnel have moved to a place called Kijevo, 11 kilometres from the Bosnian border. We are here to secure a bridge and a dam — the bridge was destroyed two nights ago by artillery fire. I'm not sure what our purpose is here in Sector South. The people down here are very poor. They are always begging for food. I'm sure lucky to live in a country with no war. I realize how lucky we are really now that I've been here. The weather here is too hot for me. Today, it reached a high of 48 degrees Celsius in the shade. Pretty hot, aye! I find it hard to sleep if I'm too hot.

I'm not sure how long we're supposed to be here, but I hear it will be [to] August 7th. I know it will be extended probably till August 14th. They keep extending everything, that's what it feels like.

On the August 26th, I will finally see my beautiful wife. She will be here for 17 days. On September 12th, she returns home. I will have approximately 15 days left when she leaves. I'm suppose to leave on September 27th for home. It will be great to see my girls again.

Well, my friend, thanks for the support. I'll try and write more.

◆   ◆   ◆

The following letter was written by a soldier who served at the headquarters of the UN Bosnia-Herzegovina Command:

Sarajevo, Bosnia-Herzegovina
August 31, 1993

Dear Mrs. Snailham,

As my time in Sarajevo and Bosnia-Herzegovina draws to a close, I wanted to say a heartfelt thank you for your kind words of encouragement over these last 11 months. You have been a true inspiration in helping to keep my spirits up as well, I'm sure for those others you may also have written to. I'm sorry that this is only my second letter to you as I have found it very difficult to find the time with all the demands which plague us here. No doubt you have picked up the reports that I am now *persona non grata* by the Bosnian government which is a disappointing way to end my tour here.

Unfortunately, in trying to maintain impartiality in explaining the situation here about the military encirclement versus the humanitarian aid "strangulation" the powers that be used the occasion to misrepresent my explanation and launched a very successful propaganda campaign to turn public opinion against the UN. *C'est le guerre!*

We can only hope that the future holds something promising for these poor people who have to endure so much. They are truly the wretched victims of this damnable war. Again you have been a real trooper. Maybe one of these days we'll have a chance to meet. God bless, and I wish you and your family the very best.

◆　　◆　　◆

The following letter was written by a corporal who served with the Signals Detachment, 2nd Battalion, PPCLI assigned to the Canadian Contingent of UNPROFOR:

Kijevo, Croatia
Postmarked, September 20, 1993

Dear Jane,

Well, I had an interesting flight back here. I spent the first part of my flight baby-sitting these two children. They were pretty funny and kept me in stitches.

The rest of the flight was good. The aircrew found out where we were going and treated us like gold. We had a really long flight to Kijevo, where we realized that vacation was over and I was definitely back to work!

Things started happening very quickly; I got promoted to corporal, spent the day packing up all my things so I could move 30 kilometres up the road, listened to the artillery coming in and started practicing for our medal parade.

The Countess Mountbatten of Burma Lady Patricia flew in on the day Zagreb had been attacked, and presented us with our UN medals at parade.

Lady Patricia is a superb woman. She is the honorary head of the regiment [our Colonel-in-Chief] . She always takes time out to come and visit the troops.

Well, it's less than three weeks now and our battalion has taken on a new objective. There has been heavy fighting in our area and the Croatians have taken over a number of small Serbian towns, after going there and butchering every living thing. It was our job to go thru these towns and sweep for survivors and non-survivors. After going through these towns we videotaped our findings for Amnesty International and other human rights organizations so that they can take action at some time or other against the aggressor.

I spent the better part of yesterday afternoon sitting in the hospital with a friend of mine that took shrapnel in his right ankle. We don't know how much damage it actually has until we get back to Canada.

Tensions are getting high here. Two of my friends were beating the shit out of each other the other day, and we were told that back in Ottawa they are under the impression that our morale is at an all time high. I don't know who told these people that but obviously they never spent five or six days in country over here with us.

A sad note, there was an accident over here, one officer is dead, one has two broken arms and one female is in critical condition in the hospital. "War is Hell"

Jane, I can't express my gratitude. The boost to my morale and the way your letter's make me smile. It is nice to know that there are some people in this country of ours that do recognize the service to world peace that we Canadians are donating, giving up our free time and careers, family and friends and love one's for something we believe in. With the support you give us and the morale boost, you are a true Canadian and it is my honour and privilege to call you my friend.

◆ ◆ ◆

The following letter was written by a 29-year-old corporal who served with the 1st Battalion, R22eR assigned to the Canadian Contingent of UNPROFOR:

**Somewhere in Croatia,
Postmarked, November 3, 1993**

Hi Mrs. Snailham,

I received your letter yesterday and I was very glad to know that somebody [was] thinking about me. I am 29 years old and have been in the force for 10 years.

For the first six years, I was in the navy [based] in Halifax. I sailed on three ships: HMCS *Fraser*, HMCS *Annapolis* and HMCS *Terra Nova*. I did two NATO tours in 1986 and 1988. I changed my career in 1990 to be [in the] military police after I was posted to CFB Valcartier in February 1992. I was posted to the 3rd Battalion, R22eR. I did my first UN tour in Cyprus in 1992 for six months. Now I am on my second tour. About my career, I have been away from home [most of] the time. My girlfriend has more problems adapting to being separate from me. About this place. I work in Croatia in the south. The camp, it is in the valley close to Gracac. The weather is the same like in Canada. The place, the country is Croatia, but the Serbs control the land. In the buffer [zone], they have a lot of fighting, but where I am it is more quiet.

I am very sorry about my mistake for my writing in English because it is my first letter I write.

◆   ◆   ◆

The following letters are three of seven letters written by a 37-year-old sergeant who served with the Public Affairs detachment assigned to the Canadian Contingent of UNPROFOR:

Daruvar, Croatia
November 9, 1993

I just arrived back in Croatia yesterday, after three glorious weeks at home. I was pleasantly surprised to have your letter waiting for me here. I was feeling lonely and homesick so your letter picked me up.

I guess some introductions are in order from my end. I'm 37 years old and spent all my youthful years in the Halifax/Dartmouth area. As a teen, I lived in the Fairmount/Springvale area adjacent to the Ashburn Golf Course. I attended Halifax West High School and graduated in 1975. I thoroughly enjoyed my teen years. Saw and did lots!

I've had an interest in the military for quite sometime. I was an Air Cadet for five years as a teen. It sure did teach me lots and prepared me somewhat for the real military. But I wouldn't say that I'm over zealous about the military. I don't sleep, breathe and eat military. Don't get me wrong, I've appreciated all the opportunities and experiences that other Canadians won't be able to fully comprehend. As well, I am proud to wear my new uniform and contribute to the peacekeeping effort. Although sometimes, it is hard to tell if we're actually providing help to the people over here. I joined the CF almost 18 years ago as a photographer. The only trade I wanted. It was (and still is) a hobby I had learned through [the] Cadets. So far, it's been a great career. Of course, [there are] a few ups and downs, but for the majority [of the time, it has been] a great career.

My first posting was in [CFB] Greenwood, in the Annapolis Valley. My high school sweetheart and I were just newlyweds, and it was a great place to start our married life and family. We lived there for five years. It was a great setting and our families in Halifax were close enough, and yet at the same time they were far enough from us. While there, our two daughters were born. In the summer of 1981, we moved to CFB Borden, Ontario where I was a photography instructor for five-and-a-half years. Our third daughter was born while we lived there.

From Borden, we were posted to [CFB] Shearwater. I guess this posting had several highlights. Most importantly, we were back home again. It's important to us as we are very family orientated and want the children to have a place they can call home. So many military families have no "roots" or place that they can call home. We also purchased our first home and lived in Eastern Passage. Work wise, I learned how to be an underwater photographer; that was something else! We had a short tour in Shearwater, only three years. Then it was off to Winnipeg where I was in charge of a 10-man photo section. We had a very trying time there as it was considerably farther from home than we would have liked. But, it was good to run a photo shop, as well as a family. We did take advantage of all the tourist attractions, saw the prairies, the USA and most of Manitoba. Except for being far from home, it was very good. We had a nice house, great neighbours and the kids enjoyed school and friends. That was only a three year posting as well.

While in Winnipeg, I found out that there would be an opening in [the] Public Affairs [branch] in Halifax, so I lobbied quite hard to win the job. It of course paid off for me, and in July of 1991, we came back home once again. I hope it's the last time we leave the province as we are so attached to this place. We live in the Dartmouth area. The girls absolutely love their friends and schools. Actually, in many ways, it's as if we never really left the area. Another critical factor is my eldest daughter is in Grade 11 and is working towards going to Dalhousie University. I would hate to (through a posting) cause the family to split up, (i.e. four out of five of us move and force one to stay behind). If she wishes to move out on her own,

that okay, but only if she wants to. So there's a brief history of me!

The situation here in Daruvar, Croatia is quite stable right now. For example, we are permitted to go into town in civilian clothing, or in uniform, eat in the restaurants, shop in the stores, etc. I'm sure we help the economy to a certain extent.

In Visoko, Bosnia it can be tense there at times and there are many more precautions taken for our safety. With Visoko and CANBATT Two being only 45 minutes from Sarajevo, there is a fair amount of tension. CANBATT One, on the western edge of Croatia is also in a "touchy area." There are daily reports of cease-fire violations, small skirmishes and of arms fire near some of our soldier's positions. I was there in September when our soldiers were caught in the middle of a firefight as they tried to force Croatian soldiers back to a prearranged line.

Thank you for writing to me. I certainly appreciate your care and concern. Too bad there's only so few that willingly write to us. Your pride in us is not wasted, as I have seen first hand what our troops are doing, and how well they are doing!

Fojnica, Croatia
November 19, 1993

Dear Jane,

Thank you for the card and letter, it sure did brighten up my day. It's comforting to know that you and other people back home are concerned about the Canadian peacekeepers.

We've had snow almost every day here for the last week. It's quite cold, probably — three to five degrees Celsius today. If it stays like this it will be a long hard winter for the poor people here. At times my heart goes out to the people, the majority of them don't want this war, they just want to resume their lives. Several of the people working here have rallied friends and family back home to send mittens scarves, etc., here to help the children. So many of the refugees have nothing. But it raises a question. Should we help them or the children at home who have nothing? It's a tough decision.

I'm by myself in my office for the next two weeks as my boss is at home on his UN leave. He's been gone for one week already. Public Affairs is a four person job to begin with, so I am definitely busy. At least the time goes faster that way.

We have been in the news lately as our troops have been trying to restore order in the Fonjnica Hospital.(Canadaian and Danish Peacekeepers had to take over Fonjnica Psychiatric Hospital and care for its patients abandoned by the stall who joined the exodus of refugees.— editor) I don't envy the peaacekeepers in Bosnia, it's a lot more dangerous there. I was there myself for two and one half months and was very happy to come to Croatia. Although here, we have some places where our troops have been in danger as well.

I am not sure what or when there will be a solution to all of this. The sooner the better, so there won't be many more people killed, especially the peacekeepers. Gotta run for now. Thanks again for caring about us!

Somewhere in Croatia
November 24, 1993

Dear Jane,

Well I guess some congratulations are in order for you! You are the first person to send me a Christmas card this year! Thank you so much. It was so nice to open the envelope and receive it! Things aren't too bad here, but if the weather doesn't change we could be in for some trouble. We are having trouble with convoys that supply the different Canadian camps. Besides the regular problems with clearances at the checkpoints and the shooting and the fighting, we are now encountering four to six days of freezing rain and snow. Some of our camps are getting very low in certain items, especially fuels. Food convoys bringing humanitarian aid to refugee camps are experiencing the same problems. The winter hit harder and sooner than people expected. But in the traditional way, I have noticed that our peacekeepers are keeping positive attitudes and are making do with what we have. This seems to be a trait that our country has over some of the other UN countries. We have much more ingenuity, adaptability — a more of a "can do" attitude.

Christmas over here will be a challenge for us who are here, away from families. It seems as we draw closer to the date, people are talking less positively about it. The general consensus is that we want it to pass by as fast as possible. At least we will be here to support each other. I'm sure we will get by. We always do!

Jane, thanks again for the letters and cards. It sure is nice to hear from you (actually, read from you)!

◆　◆　◆

The following letter is the first of two letters written by a 38-year-old lieutenant-colonel who commanded 12 RBC assigned to the Canadian Contingent of UNPROFOR:

Visoko, Bosnia-Herzegovina
November 28, 1993

Dear Jane,

If you like, I'll tell you a little of myself; what our mission is in war-torn central Bosnia; and try to describe the living conditions of the people in our area of responsibility (AOR).

I'm 38 years old; am married to a beautiful French-Canadian woman and I have two children, aged eight and five respectively. When I'm not vacationing at my winter retreat in Bosnia, I live in Valcartier, just north of Quebec City. Like you, I also come from a small town; in this case, from Northern Ontario.

Our principal mission is to escort and to provide armed security for the humanitarian relief convoys (food, fuel, etc.) which are operating throughout Central Bosnia. Our secondary mission is to provide security for the UNPA which surrounds the Muslim enclave centered on the village of Srebrenica, in Northeast Bosnia. There are presently plus or minus 20,000 refugees trapped in the enclave, surrounded by the Serbs. So far we've had a few close calls, but no one has been injured. I hope I can keep it that way.

The common people of Bosnia live in appalling conditions. The problems are further compounded by the early arrival of winter. Imagine your neighborhood.

Now imagine every home in your neighborhood without heat, electricity and hot and cold running water. None of the houses have glass in the windows. Imagine groups of armed men roaming your neighborhood at night, stealing your food and threatening to kill everyone, or rape the girls and women. This is the sad reality of this country.

Rest assured, we will continue to make the difficult decisions whatever the consequences. Thanks again for your letters. Please convey my warmest regards to your husband and son. Merry Christmas and a Happy New Year.

◆ ◆ ◆

The following four letters are the last of seven written by the 37-year-old sergeant who was mentioned earlier on pages 31-34, and who served with the Public Affairs detachment assigned to the Canadian Contingent of UNPROFOR:

Daruvar, Croatia
November 30, 1993

Hello pen-friend,

Hi Jane. Wow! You are flooding me with mail! I'm so excited to receive mail, and through my family and you, it's been super! I can't say there's been to much going on for me lately. My superior is away on his vacation still, so I'm staying quite close to the office. I guess the biggest news of the last two days, I say big news, not good news, is that we lost two more of our guys over here. Jane, the rumours and bad news travels like "wildfire," but at the same time you can easily detect the sullenness and the deep regret we all have. "Did you know him? What happened?" The questions, the questions! It's always such a tragedy to lose peacekeepers. It's as if a part of everyone of us over here has died. If these two young men have wives and children, as well as parents of course, and loved ones, my heart goes out to them. I guess so many of us here in our own different way will be affected by this loss, and silently we'll mourn. Is it necessary, do we really belong here? Are we doing the right things? It sure does put doubts in your mind — enough of that.

I hear you too are a TV celebrity, congratulations! Good show, what you are doing for me and other service people shouldn't go unnoticed. As you've told me before, it's you way to contribute and show you care. Briefly, I'm really looking forward to seeing the Christmas card! Sounds great, thanks in advance! Sorry, I missed you on the show yesterday. Thanks for the angel, it's on my lapel of my coat. I've gotten numerous comments about it! Your Christmas card was the first one that I received this year!

Thanks so much for taking the time to write to me. I really do appreciate just how much you care!

———————————   ◆   ———————————

Daruvar, Croatia
December 12, 1993

Dear Jane,

I received your huge Christmas card today! Thanks it was great. It was nice to read some of the comments that people wrote. I'm glad to see we have such support. For the 1,500 of us over here during the holidays, any vote of encouragement is a positive motivator for us. It's going to be tough, no question about it. I'll be so happy once Christmas is past. I received a parcel from my wife as well today. It sure did put a lump in my throat and a weight on my heart. It's funny. She says, "Don't worry about us, we'll be okay, just make sure you are okay." I then say, "Don't worry about me, I'll be okay, I'm worried about you guys." So we will either worry about each other, or hopefully, not worry at all. It will be very strange and very difficult to be away from my girls, but I know that we all will be okay, as we are [a] strong and close family. I'll be back home at the end of January, so we'll celebrate Christmas then.

I was videotaping the camp choir as they practiced tonight. At the end, several people came up to me and asked if they sounded good or not. Jane, all I could say to them was as long as their songs came from the heart, then it would sound beautiful! They were practicing my favorite Christmas song, "O Holy Night," so I was not very impartial. Gotta run for now, but I'll be writing again! Thanks again for the "mega" card as well. Thanks for your concern. It goes a long way. "Happy Holidays" to you and your family. Your contribution to peacekeeping helps make what we are doing worthwhile. God bless.

Daruvar, Croatia
December 23, 1993

Hi pen-friend,

The public affairs officer (PAO) and myself just came back from a four day tour of Sector South that our boys are occupying. There are a couple of tense spots and they are regularly shelled and shot at. It's not a fun task. Other parts of the sector are relatively calm. We were there to set up some radio interviews. We also set up some photo opportunities of peacekeepers working through the Christmas period. We had great success, approximately 30 interviews and four photos sent on the wire service.

The morale is pretty good considering the time of year and the circumstances that a lot of these young soldiers have been placed in. Over the next few days there will be a few small celebrations, but for the majority of the people, it is business as usual. For me, I'll be patient, as I'll be home in a month, and we will celebrate then!

I think it would be a lot harder on me emotionally if I had three to four months to go. But, my time is almost over, and I'll be so glad to be home with my family and friends.

Yes, I like sports very much. I try to be as active as I can. I'm willing to try almost any sport. I wouldn't say I'm star status, but I can perform above average in most of them. But when I'm surrounded by four girls, watching a hockey game can be a challenge, actually it's dangerous for my health to take the remote control away from them! I didn't follow the regular season too much, but I (as does my wife) enjoy the play-offs. Well, it's time to get my teeth brushed and off to bed, 0630 comes early.   Thanks again for your letters and support! It sure is nice to hear from you!

<hr>

Daruvar, Croatia
January 16, 1994

Dear Jane,

I've spent the last three-and-a-half months in a Croatian environment. Two friends (actually families) I know are Serbians living in a Croatian community. Yes, they did what they had to do to survive, but through the war, they remained loyal to their towns and neighbors, eventhough they were (and still are) in fear of death.

Like many people in Canada and throughout the world, we hear of the aggression of the Serbs and the plight of the Croatians. But Jane, I've been in many parts of this country [in] areas won [and] lost by all three types of people: Serb, Muslim and Croatian. I can assure you that at the lowest level, people are people, humans who want the war to stop and life to carry on. And at the same time, I've seen it at higher levels, where given the chance, one army is liable to be so diabolic as the one it opposes. Sorry Jane, it's not just the Serbs, it's everybody. The war has affected so many people in so many ways. Am I making any sense to you, or am I just rambling? I must admit that at first I felt the Serbs were the "bad guys." But after what I witnessed in Medak [on] September 18, 1993, I know it's not true.

Well enough gloom and doom! I had an opportunity to photograph and videotape one of the padres and soldiers distribute food and clothing to the people the other day. It was wonderful to see the donations that came from the Canadians. It was as well, gratifying to see the people thankful to receive boots, coats and warm sweaters. It's small gestures like that, that restore my faith in people; the caring that we, our fellow Canadians did for the people. It was great that my photos made it into at least six major newspapers and Maclean's magazine. The videos were shown on two networks and at least three radio interviews were made. Good mileage for our troops. It's nice to see a happy story once in awhile.

I've started to prepare for my return home next week. I've done some advanced packing of relevant kit and it's to be forwarded ahead of me. I have also slowed down a little — not taking on too many big jobs, and I've reduced my work days from 16 to 12 hours. So, I finally have a bit of time for myself.

Throughout this tour, I've made a lot of wonderful friends. Military, yes! We all depend upon each other, so you make friends fast and you learn to trust each other quickly. The military is small, and the chances of seeing some of the people again are good. On the other hand, the people of Daruvar I've met, I'll probably never see again in my life. That makes me sad, and I also worry about their futures. There's a couple of translators that I know, one of their families as well. Also, the family that owns the local photo lab have been wonderfully kind to me. For all of them, their friendships mean a lot to me. I'll cherish the time I've had with them forever! I get pretty sentimental about things like that. I know as well, when I say goodbye to them; it will be very emotional.

I have, through my experiences, gained a lot of self-confidence as a person, soldier and photographer. I feel that photographically, I'm probably at my best right now. All the years of training, practice and knowledge came together and through the images I've created, I know that I am a very good photographer. As a person, I know that I've matured. I see life in a different manner for the better. I'm so thankful for what I have, and I'm especially thankful for "who" I have. My wife and the girls have been my strength. She supported me all the way. Her love was the one constant thought [throughout] the entire six months! Militarily wise, I've learned to take control of situations, and make decisions on my own. Moreso, than I've done in years. We were passing through a checkpoint and a junior officer on a media tour we were escorting got caught sticking a camera out of the window. The way I handled the situation, negotiating with the Croatian army, ensured the safety of my passengers, was such a "rush." I really was in control of the entire situation on my side, and it was a great boost for my confidence. Too bad it made the officer look like the idiot, but, oh well!

Jane, thank you for all the support that you have given to me. Your cards and letters always perked me up, and I'm so happy to know that there is someone like you at home that cares so much about peacekeepers throughout the world. You are a true inspiration. Too bad people ridicule you rather than follow you! Perhaps someday they will understand! But in the meantime, please keep writing to whomever you can, whenever you can. To know that you care, makes it just a little bit more rewarding for the peacekeepers in their missions. Thanks again, the support has been wonderful. See you back home

◆ ◆ ◆

The following letter is the second of two written by the 38-year-old lieutenant-colonel who was mentioned earlier on page 34, and who commanded the 12RBC assigned to the Canadian Contingent of UNPROFOR:

Visoko, Bosnia-Herzegovina
February 12, 1994

Dear Jane

Well, the "banditos" have finally granted me a respite so I can reply to all the wonderful cards and letters you've sent me over the course of the last two months. Please excuse the tardiness of this letter, but the psychotics who inhabit my AOR have been very busy little boys lately.

The beginning of February heralded the halfway point of our duty in Bosnia. As you can well understand, all the "horses" can now smell the barn; however, I've had to tighten the disciplinary reins in order to ensure that no one forgets where they are and the danger that lurks beyond the defensive perimeter. To this point in time, I'm satisfied that we've done everything we can to reduce the levels of misery and hardship experienced by the people. So far, we've escorted 80 humanitarian aid convoys, delivering approximately 5,050 tonnes of food and over 100,000 litres of fuel. To do this, we've had to skirt running battles between the warring factions. Unfortunately, we've had two men killed, four wounded and two trucks have been badly damaged by machine gun and artillery fire.

I very recently returned from a visit to Muslim enclave of Srebrenica in Northeast Bosnia. Notwithstanding, the marvellous job done by the 160 women and men I have stationed in that particular area, there is only so much that we can do. The living conditions of the 44,000 people trapped in the pocket are atrocious. The hills surrounding the town are a moonscape, having been totally denuded of all foliage, which is being burned to produce heat. Piles of rotting garbage picked clean by the inhabitants, are infested with rats and other vermin. There are thousands of streetwise, orphaned children, their parents having been killed in the fighting. Srebrenica has been machine gunned, bombarded and mortared back to the last century.

The last 30 kilometres of the road we use to reach Srebrenica bears witness to the full fury of the wave of hatred and ethnic cleansing which washed over the area last spring and summer. All of the 15 or so villages and hamlets which lie astride the east-west road have been totally destroyed. There are no people, no animals, no stray dogs or cats, even the birds have gone. In the fields, are low piles of dirt. One can only imagine what lies buried beneath them. These images will forever remain indelibly etched in my mind and those of my women and men.

Thank you once again for providing me with a link to the outside world. The photograph you sent me enables me to put a face on the name. Please convey my warmest regards to you husband and son.

The following two letters were written by a captain who served as a helicopter pilot aboard a replenishment ship assigned to Operation Sharp Guard:

HMCS *Preserver*,
on patrol somewhere in the Adriatic
February 14, 1994

Dear Jane,

I have learned that when you are so far away, a letter from home, even if it's from someone you haven't met yet, can sure brighten up your day. Thank you!

So to introduce myself, I was born in Amherst, Nova Scotia (NS), but my parents moved to Ottawa when I was six months [old]. I lived there until I joined the military in 1989. I like to joke that I joined the military completely ignorant. I thought basic training would be fun and pilot training a blast. Well, I was wrong on both counts; it was hard work from the word "go."

But, there is good in everything. I met my husband in Portage la Prairie, Manitoba. I wouldn't call it love at first sight, my mind was on the job but there was an overpowering force. He proposed in a romantic evening on the first anniversary of our first date. We had a long engagement because I wanted to be sure that we would be posted together. He is now posted to CFB Greenwood and I am in Shearwater. We have a house in Windsor, NS which is exactly the halfway point.

We have been separated before, but never for so long. I've never gone so long without talking to him either; but, everyone on the ship is in the same boat (pardon the pun) and we are all supporting one another. I've never been overseas, so there is some degree of excitement. Today, we are going into port. We are going to the island of Corfu, Greece. HMCS *Preserver* is a supply ship. While we are in the Adriatic Sea, we replenish the other ships. So every three or four days, we are back in port, but usually just overnight. The helicopters are being used as a taxi service. There are senior officers from all nations who need to get here and there. It has been interesting because it is so new. I can see it becoming a little dull, but we are support and I guess that's what support does.

The countdown has begun. There are 121 days and wakey, wakey before we sail into Halifax harbour again. Wakey, Wakey is the whistle they blow in the morning to wake those of us up that have the privilege of sleeping through the night. Every Friday, our soup of the day is clam chowder. Sixteen more chowders to go!

Thanks again for writing, Jane, and Happy St. Valentine's Day

HMCS *Preserver*,
on patrol somewhere in the Adriatic
March 23, 1994

Dear Jane,

It's amazing how the ship buzzes with excitement when we hear there is mail on the way. So I share your letter with those who are not as lucky.

Apparently, we have 84 days and a wakey, wakey, 12 more chowders and April 7th will be hump day [meaning the ship is halfway through the deployment]. The best part, however, is that the ship is going to go alongside for three weeks starting April 11th. That would allow us to take leave, and my husband will be flying out to meet me. We are planning to celebrate our first anniversary a little early in Paris and Austria. I guess there's a positive for every negative.

Life has become pretty routine. Same faces, same place. But good friends are being made and it's becoming a more comfortable place to live. Now that I feel more comfortable, I realize how awkward it had been at first, 340 strangers living together in such close quarters.

Just a few days ago, we sailed from our last port visit in Corfu, Greece. It was a disaster. Eight people were injured in motorcycle accidents. The rules out here are very lenient. Fortunately, no one was seriously injured. A friend and I, instead of renting a motorcycle, rented a room at the Corfu Hilton. Our room overlooked Corfu International Runway, so we shared a bottle of wine and watched the planes come and go. Always a pilot, I guess. And as soon as the giggles started to go away, I phoned home and spoke for two-and-a-half hours. I'm grateful I won't see that phone bill. It felt as though I was sitting at the kitchen table with my husband, discussing the events of the day, like we used to do. It was great, an excellent port visit for me.

The flying has picked up quite a bit now. Keeps us busy. Because we are on a supply ship, we do a lot of passenger transfers, mail and cargo delivery and [shuttle] dignitaries. We spend a lot of time [flying] over ships from other nations. They have also been tasking us to enforce the embargo and to find and hail merchant cargo vessels. The flying time is doing wonders for my confidence level. The weather has been great as well. Not too cold, not too warm and no storms. Always flyable.

Tonight, the ship has organized a casino night, in support of [the charity] Big Brothers, Big Sisters. The prizes have all been donated by the ship's company, like the cooks have offered up a candlelight dinner for six, the liquid control officer has offered a case of beer, and breakfast in bed the next day. What a great idea, it breaks the monotony, brings us closer together and it's for a good cause. I always wanted to be a Big Sister, but I never seem to be in one place long enough.

Anyhow, thanks again Jane, for your letters and I promise to keep smiling!

⬢   ⬢   ⬢

The following letter was written by a 29-year-old master-corporal who served with Alpha Company, the 1st Battalion, PPCLI assigned to the Canadian Contingent of UNPROFOR:

Somewhere in Croatia
April 30, 1994

Dear Jane,

I have to apologize for the delay on this letter, but there is so much work for me here in the evening, I just want to crash! It's 0100 here now, so I'm taking a few quick minutes to write as I do for letter [from] home, too.

So a little bit about myself. I am 29 years old and married to a wonderful civilian husband for four years on May 19th. I have two wonderful sons. It was very difficult to leave them and I think about all my three boys everyday. My husband and I were prepared for this tour, as we expected it for about a year. We decided to make this tour work for us and save as much money possible to buy a house. So it's a difficult time for better things for all of us later.

I have been in the armed forces for nine years now and still enjoy it as much as the first day I became an army cadet. I am a vehicle technician, and here in the former Yugoslavia I am responsible for 16 tracked vehicles and 12 wheeled vehicles. I am attached to A Company, a rifle company of infanteers of the 1st Battalion, PPCLI in Calgary.

I arrived in Calgary in May 1993 from a three-year tour in Germany. I would go back there in a minute. My husband and the boys will meet me there for my UN leave in September. Our friends in Germany have built a room for us on their house and are very anxious to see us.

About where I am and what I'm doing here. I am in Karin Slana, a bay off the coast of the Adriatic Sea. It is very beautiful. We have taken over some OPs from the Croats and the Serbs and we man them 24 hours a day, seven days a week. We report anything suspicious to [our] higher [headquarters] and are beginning to confiscate weapons from the persons inside the ZOS. My specific job, of course, is to repair anything that breaks down.

The saying "War is Hell" is quite right. The houses close to the line of separation are totally destroyed. It's very sad. The people have no work, the children are hungry and are raggedly clothed. Some wave excitedly when we pass by and others show us the finger or the victory to Croatia sign and the victory to Serbia [sign]. It's hard when they run to the street making signs of hunger, putting their fingers to their mouths. I try to take some fruit and cookies to them, but sometimes there's not enough for us. Anyway, I try to stay away from the sad parts of this war, it gets me down. I'm just going home appreciating Canada. Thanks for your letter Jane.

●　　●　　●

The following letter is the first of three written by a captain who served with Reconnaissance Platoon B of the 1st Battalion, PPCLI assigned to the Canadian Contingent of UNPROFOR:

Benkovac, Croatia
May 4, 1994

Dear Mrs. Snailham,

I am an officer in the Canadian army writing to you on behalf of my troops: [the] Reconnaissance Platoon, 1st Battalion, PPCLI. Your letter to "a Canadian peacekeeper" arrived early in our tour and I posted it for all to read. Many of the troops were pleasantly surprised that someone who didn't know them would take the time to write such a letter. I can tell you honestly that we realize that many Canadians do not appreciate, far less have an understanding of, what we are doing or why we are here. Of course it is not an easy issue, but I guess in a nut shell, we see ourselves as trying to make things better in a very messed up part of the world. On a more personal level, I think we see the little children that run out to wave as we go by, and we think of the kids at home and how fortunate they are and how sad it is for these ones, to live in a country where the adults wreak mindless destruction.

Well, enough of that! Since our arrival, we have been very busy implementing a cease-fire between the Serbs and Croats. My troops man OPs in the ZOS and conduct vehicle and foot patrols to maintain a UN presence in the area. When we have time to relax, we make good use of it, catching up on lost sleep, writing the odd letter and generally just sorting ourselves out for our next job.

We are a 31-man platoon living in a camp near the town of Benkovac. We live in

25-foot-by-eight-foot trailers, which are not too uncomfortable, compared to tents! The weather is starting to get hot. Presently, it hovers about minus 20 degrees Celsius midday and we expect summer temperatures in the area of 30 to 35 degrees Celsius as the norm. Walking around in helmet and fragmentation vest under such conditions can be trying at best, and extremely uncomfortable, irritating and wet, at worst.

Recce Platoon B (as we are known) is hoping to get a dog in the next couple of weeks. No doubt it will be a good morale booster for all the troops. Being away from family is the hardest part and it is nice to come back to a devoted animal. Well, that's all for now. Best wishes from all and thanks again for thinking of us.

●　　●　　●

The following letter is the last of three written by the soldier who was mentioned earlier on page 16, and who served with the Administration Company of the 3rd Battalion, PPCLI, but was transferred to the Headquarters Company of the reconnaissance element of the PPCLI assigned to the Canadian Contingent of UNPROFOR:

G. Mirange, Croatia
Postmarked, May 29, 1994

Mrs. Snailham,

Again, I appreciate you sending the letters and cards, the angel I put on the inside collar of my combat shirt. My mom is very religious. Unfortunately, I never did get around to going to church very much, however; I figure any extra power I get from the "big guy" couldn't hurt. I plan on wearing that angel, everytime I leave the compound, which is a lot. Now I'll answer your questions.   We are at G.Mirange which is about four kilometres south of a larger town called Benkovac. These towns are all in Southern Croatia.

Recce platoon is the British term, the Americans call us a recon platoon. We are mostly employed as scouts. We usually go into areas first and navigate the routes, observe and report back to the following units anything we see, i.e. gun emplacements, belligerent patrols or activities.

Usually when we patrol an area, it would be at night with night goggles, and myself and two or three other men. We would move in on foot, through woods or bush and observe an area that's of particular interest to the battalion commander, however in UN operations we are supposed to use roads and let the people see us. So we have to adapt a little to UN operations, here we want to stay on the clear routes, too many minefields to chance going off them.   Within the unit, we are called the "eyes and ears" of the battalion. It's hard to tell anyone what I do because there are so many different things we can be tasked to do. But, what I told you is our basic use, we're just not allowed to be "too sneaky" in UN operations. It would spook the belligerents too much. Anyway Jane, thank you very much and I'll write you again.

◆   ◆   ◆

The following letter is the second of three written by the captain who was mentioned earlier on page 43, and who served with Reconnaissance Platoon B of the 1st Battalion, PPCLI assigned to the Canadian Contingent of UNPROFOR:

Benkovac, Croatia
June 19, 1994

Dear Jane,

You obviously realize how important it is to get mail while away from those you care about. I look forward to letters from my wife, especially those with pictures of my one-year-old son. When I left Canada, he was 11 months old and still looked like a baby, now he is almost 14 months and he looks like a little boy. Imagine, how quickly he is growing up. I only wish I could be home to see more of it.

Well things are settling in a bit of a routine finally. Although we still have the little flare-ups, most of our time is spent dealing with investigating small arms shooting, mine explosions and the belligerents entering the ZOS. My platoon does a lot of patrolling, both on foot and in vehicles. It makes for long days under the hot sun!

The troops morale is pretty good, despite the uncooperative nature of the sides we deal with.

Most of us took some time out to watch the D-Day festivities. It makes you wonder though, how most of the talk was about the British and American forces and their sacrifices, almost as if we were third rate players, especially when we all know how large and important the Canadian contribution was! Fifty years later and we still can't get the credit we deserve — go figure. Take care.

◆　◆　◆

The following was written by a Canadian soldier who served as a PAO assigned to the Canadian Contingent of UNPROFOR:

Visoko, Bosnia-Herzegovina
July 21, 1994

Dear Jane,

We've been very busy here. As you may have heard, on the night of July 3rd, there was a firefight between the Serbs and Muslims around one of our OPs. Our mortar troops had to fire over a dozen 81-milimetre illumination rounds to keep things quiet. This was the first time [Canadians] fired "in anger" since the Korean War.

Unfortunately, in a separate incident that same night one of our young armoured car crew commanders was slightly injured when a piece from a bullet hit him in the cheek. At the same spot, on the night of July 14th, our Canadians had to shoot at both the Serbs and Muslims in order to withdraw from the OP. Apparently, it made the news back home. Things have been quiet since then, and as of yesterday, we have reoccupied that OP position.

We've just heard that the Serbs have "conditionally" agreed to the latest peace plan. We await further news on this. Don't know what the implications are; whether we'll stay or get pulled out. Many of us want to stay.

As the PAO, I do move around a lot. I'm a "crisis communicator" and handle all media queries when something happens. I also write the news releases which ultimately make it into the papers at home. The days are long, but never boring. Got to go. Take care, write soon.

◆   ◆   ◆

The following letter is the last of three written by the captain who was mentioned earlier on pages 43 and 44, and who served with Reconnaissance Platoon B of the 1st Battalion, PPCLI assigned to the Canadian Contingent of UNPROFOR:

Benkovac, Croatia
July 26, 1994

Dear Jane,

Well, it's been awhile since I last wrote...sorry. Things have been just cruising along here and the problems are on the rise, just like the temperatures. It's been hovering between 30-35 degrees Celsius for the past month.

The Serbs are really and truly becoming a royal pain in the behind! They don't keep their word, they steal, lie and generally impede our actions every chance they get. It is becoming more and more difficult to do the job we were originally tasked for. The Croats are slowly catching up though as I can attest. They have now closed all the entry points into the UN area of operations, effectively cutting our system of overland resupply. I was guiding a convoy with one of my sections of troops [a section contains eight soldiers] this past week. We were supposed to go to our new battalion headquarters in Rastevic. The convoy was 47 trucks filled with food, fuel and our mail. The Croatians stopped us at a bridge and refused to let us pass. We spent five days in a stand off. In the end, the convoy turned around and returned to Split [located] on the coast. It was very tense the very first day as the Croats made all sorts of threats and on one occasion, even cocked a weapon. Of course, as Canadians we told them to "bugger-off" and we were ready to shoot it out. Fortunately, it didn't get to that point, but I can tell you, many of my soldiers are frustrated; they are just waiting for either side to provoke us in the legal sense!

Besides the stupidity of this place, we are not doing too badly. Most of us were upset when one of our own was killed. Of course, the engineers were hit hard by this accident, but my troops, who were working with the engineers for the clearance did not escape unscathed, both physically and emotionally. One of my sergeants received a bad shrapnel wound in his arm, but managed to give vital first-aid to the engineer sergeant who had gotten a bad chest wound. Some of my other troops had to provide first-aid to him to try and keep him alive, unfortunately, to no avail. Having gone through a similar situation in early April, I know how hard it is for the troops to see a comrade so badly injured and then hear he didn't make it.

On a happier note, we have five or six dogs hanging around our camp now. They are great for morale, especially when you come home from a long patrol or get caught up in a convoy and as soon as you dismount from your vehicle, the dogs start running towards you and whining and wagging their tail as much as it is worth. It's almost funny! Thanks again for writing, though I probably won't be getting much mail through the Croat blockades in the near future. It's always nice to know there are people back home that are interested in what we are doing and how we are keeping.

The following three letters were written by a captain who served with the Combat Support Squadron of the RCD assigned to the Canadian Contingent of UNPROFOR:

Somewhere in Bosnia-Herzegovina
December 24, 1994

Dear Jane,

Merry late Christmas. I received your other letter and was about to write back when I realized that I never kept the envelope with the address. Thanks for the picture and your cards.

No, I was not detained by the Serbs, but I sure filled and carried a lot of sandbags. As you know, the incident was resolved. My squadron has now replaced the troops that were over there.

So far, the Christmas season has not been too rough. I just got off the phone with my wife and kids. I got them out of bed at 0515 as my allotted phone time is 1115 here. My in-laws are down from Sasketoon and they're going to a friends house in the country for Christmas dinner. I'm pretty busy here writing letters and on duty: radio watch. I'm hoping to catch up on the reading. A number of packages I've received from home feel like books.

Things are looking up here, in 10 minutes, [former US President] Jimmy Carter's cease-fire goes into effect. For the past week, the number of shooting incidents have been declining. This is a change from other plans where the fighting has increased as the deadline approached. Hopefully, it will be "two and zero" for Carter. First Haiti and now here. Maybe he should be the next Secretary-General of the UN?

From my family to yours Jane, I hope you have had a splendid season. For next year, I hope for you the best. Thanks for your support.

Visoko, Bosnia-Herzegovina
February 26, 1995

Hi Jane,

I just got back a short while ago from three weeks in Canada and found these letters waiting for me. It sure was great being back with the wife and the girls. It sure was difficult leaving to come here again. We bussed from Petawawa to Toronto, [then] a plane to Rome, [followed by a] bus to the coast, [a] ferry to Croatia and [then] a bus to Visoko. Two-and-a half days! What a trip, and I picked up a wicked cold just before leaving. Sounds like you had the same thing.

Not much going on in our area, we get the feeling that things are building up for the big one. The weather is improving rapidly with days of up to 16 degrees Celsius. The general feeling is that soon the fighting will start up again.

It's is a shame about the airborne [regiment] disbanding. It sounds like a decision based on emotion versus careful thought. Most of the persons involved [in the Somalia incident] are probably not in the regiment anymore.

You're definitely right. The more I travel and see the world, the more I don't want to leave home. We have it so good yet, people want to wrench our country in two. I'm glad that almost every unit of the Canadian army has seen the effects of civil war. Let's not ever do this at home.

So, it was you who did the giant post card. Thanks so much. I talked to the padre who also sent you a letter, I believe. It's great knowing people care. Thanks for the invitation to Halifax. I don't know when I'll ever be out that way. There's not too many army units there. Thanks for writing.

——————— ◆ ———————

Visoko, Bosnia-Herzegovina
April 30, 1995

Hi Jane,
This will be my last letter. Thank you so much for your unflagging support. Its always great to get mail. This afternoon the Vandoos [the nickname of the R22eR named after vingt-deux which means 22 in French] start to arrive. Yesterday, our first flight left. I'm on flight three. In 10 days, I'll be home.

It's a good idea about your book. Best of luck with it. Of course, you can use extracts or whatever from my letters. I find the media is not too interested in items unless they are really exciting. That is to say they have to sell newspapers or give them great ratings. I can understand that because running the news is a business and it has to make money in order to survive. Most of the stuff, I've personally done has been very boring with about three hours of real excitement over the [past] seven months. That's not to say that we as a group have not made significant contributions, its just that my particular job this time around made it hard to keep my eyes open.

I'll leave you a name of one of the new guys coming in to replace us. He is [a] Major. He was here earlier this year to get a feel for things prior to his unit's deployment. I sure look forward to seeing him again. I'm sure he'd appreciate [a] letter. Again, thanks so much for your thoughts and prayers. Good luck with your book.

◆   ◆   ◆

The following letters are two of three written by a major who was stationed at the headquarters of the Canadian Contingent assigned to the UN Peace Forces (UNPF):

Zagreb, Croatia
October 20, 1995

Dear Jane,

You're sweet! I got your letter, card and angel. In fact, when I was interviewed on the English service of *Radio Canada International* (RCI) last week, we spoke about your "guardian angel." I told the interviewer you gave me an angel and I told her I had it on my right shoulder under my rank, and I always have it with me and show it off to a lot of people. Thank you.

I have been very busy since I arrived in-theatre. First of all, getting adjusted to the new time zone took every bit of a week and also, I had to learn my new job, which involved a bit of travel.

When I arrived here, we flew in from Calgary, Alberta — where I had to go to meet everyone else who was deploying with me, so I had to take a UN chartered aircraft from Calgary to Split, Croatia. Even after eight hours in the air, I found Split beautiful. We came in on a long, low descent and I was able to see all the city. [I saw] white houses with red tile roofs — some alone in fields, others built into cliffs — actually using the cliff face as one of the walls.

When we landed, we were bussed to Primošten, Croatia where we have [the] CANLOGBATT. Here, we were given our reality check. The beauty of the region notwithstanding, we were issued with flak vests, helmets and rifles. The country is so beautiful, but is being destroyed with politics.

As I write this, I'm listening to a tape of some music I have at home. Right now it's playing Bach, and it's so relaxing, that I'm really mellowing!

The job here is quite busy and with the repatriation taking place, I expect we're going to be even busier. I've developed a newsletter for our soldiers to keep them informed about what's happening. Everyone hopes to be home for Christmas. I hope they're not disappointed. As for me, I expect I'll be staying until next spring. If Canada agrees to participate in a NATO operation here, then I expect I'll be part of the operation. I brought my green beret, just in case.

Zagreb is a beautiful city with antiquity mixed in with the old, but not ancient. The city square is always filled with people and the movie theatres play American movies with Croatian subtitles. All in all, it's not too bad. I do have to go into some of the affected sectors and into Sarajevo from time to time. So far, things have been very quiet wherever I go, but I do know there's a fair bit of action in the Krajina [sector] escorts. I "see red" when some of the soldiers here try to blame the UN for their problems. When I was in a convoy last week waiting to get into the Canadian camp at Visoko, two soldiers from the Bosnian army yelled "shame" at me, I shot back with "f— you." The only trouble was I had a mini-bus full of reporters and I had to explain that these people, not the UN or Canada's contingent are responsible for the war and the atrocities that go with it The UN doesn't have the personnel or equipment to stop the war. Only by its presence ant its vigilance can it expect the belligerents to stop fighting.

Heck Jane, it's a complex mess and the animosities go back to the Ottoman and Austro-Hungarian Empires. We'll see if cease-fire number 35 works! Many thanks and keep writing.

———————————   ◆   ———————————

Zagreb, Croatia
November 15, 1995

Dear Jane,

Your guardian angel is getting quite a work out these days and I truly believe you gave me a very protective angel. Last week, I went to Sarajevo and met some people. On the way back to Visoko — where I arranged for an escort of some journalists, two minutes behind me was a minivan with some Norwegian soldiers who came under direct fire. I don't know why them and not us, but only one [Norwegian] was slightly injured.

Then last week, our camp at Zagreb Airport, they found two mines in a piece of ground that I must have walked over at least three dozen times. I got quite a startle when I realized what happened.

So I still wear your guardian angel on my right shoulder under my epaulette. Life has been very busy here since we started our repatriation. It looks like all Canadians will be home by February 1996, and I expect to be home by Christmas. After New Years, it will be back to work for me.

This week, we have the signing of the peace treaty for Bosnia and a peace agreement for Sector East [in Eastern Slavonia]. So it appears for now, a lot of animosity for people to get over and it will take a lot of strength to do so. I hope they are up to it.

With the peace accord and the withdrawal of the UN, it looks like NATO will be brought in to replace the UN [and] to supervise the peace implementation process. If so, then I might be brought back or left here as an advance party for incoming Canadians. But first, the Canadian government has to decide if we're to participate and then what we'll contribute. That may well take decisionmaking powers that are absent in many of our politicians.

Well, it's midnight, and my bed beckons and calls. I look forward to your next letter.

◆   ◆   ◆

The following letter is the first of four written by a naval officer who commanded a frigate assigned to Operation Sharp Guard:

**HMCS *Fredericton*,
on patrol, somewhere in the Adriatic
December 31, 1995**

Dear Jane and family,

Hello and thank you ever-so-much for your card and kind thoughts. It is the sincere best wishes of Canadians such as yourself that makes it a little bit easier to be away from home.

I must warn you, our mail service is somewhat slow, as it depends on military air freight schedules and our proximity to the base when it is collected for distribution to the forces at sea.

About myself? I'm from Ontario originally, but have lived all over [Canada], both before and after joining the navy. I joined the military in 1971 at College Militaire Royal de St. Jean and graduated with a Business Administration degree in 1976. Since then, I have served on several ships and have had several shore appointments in Vancouver, Toronto, Ottawa [all three are in Canada] and Norfolk, [Virginia, USA]. I guess the highlights of my career were my appointment to the Order of Military Merit (the military's equivalent to the Order of Canada) in 1991, my tour as the Commissioning Executive Officer (second in command) onboard HMCS *Halifax*, (the first of our new patrol frigates) and obviously, my pride and joy, as the Captain of HMCS *Fredericton*.

I'm married to a wonderful Montreal girl and we have two great teenagers, plus two big dogs. This is actually the second Christmas I have spent away from home in the past five years. In 1990, I was deployed with the Canadian Task Group to the Persian Gulf in Operation *Friction*. This doesn't make it any easier, but at least I've been through it before, whereas about 90 per cent of my crew has not.

We did have a good Christmas, nonetheless. We were in Trieste, Italy with a Portuguese [ship] and [a] British ship, so there was a good opportunity to get together with them. Christmas Day, however, was strictly a FREDDIE (that's our affectionate name for the ship) affair. We had gift exchanges, a traditional Christmas dinner where the officers served the troops, followed by social [gatherings] in the three messes. The place was packed and a good time was had by all.

During the visit to Trieste, we also gave the International Red Cross all the clothes we collected for Bosnia. We loaded over a tonne of clothes into a truck bound for Sarajevo.

I should take a moment to talk about our operation. We are deployed in Operation *Sharp Guard* which is the UN-sponsored embargo against the warring factions in the former Yugoslavia. This operation has been most successful and no doubt significantly contributed to the signing of the DPA. With that signing, operations are shifting to provide a mix of patrol and training opportunities. Thus far, Freddie has interrupted 70 ships, boarded 16 and diverted two to Italian ports. So Freddie is doing the job it was sent to do and performing superbly.

Well Jane, it's late and you never can tell what the night will bring. This is a bit of an odd way to spend New Years, however, because of us and those who have gone before, I'm sure 1996 will be a much better year in the former Yugoslavia.

Have a very Happy New Year and best wishes for 1996.

◆ ◆ ◆

The following letter was written by a 25-year-old corporal who served with Golf Company, the 2nd Battalion, RCR assigned to the Canadian Contingent of IFOR:
Somewhere in Bosnia-Herzegovina
January 26, 1996

Jane Snailham,

Thank you for writing. It is nice to know we have people in Canada who support us. I have only been in the army for six years, but it seemed to me that most Canadians did not support the army.

To tell you about myself, well where to start! I am 25 years old and married to a beautiful and supportive wife. I have a 10-month-old son, who I miss greatly. We live in Oromocto, NB, just outside CFB Gagetown.

I am presently posted to the 2nd Battalion of the RCR. I am an electric and optic technician. I fix computers, missile guidance systems, all optical sighting systems and I am also the electrician who wires the camp and provides heat and lights. So I am usually a very busy man, which is good, it makes time go by faster.

I am from Newfoundland. Saint David's to be more exact. It is a small outpost on the west coast about half way between Port-aux-Basque and Stephenville.

I enjoy my job and I love the army, but it is hard on the family. I am missing all the first steps, bumps and words that my son has. My wife is great.   She supports me fully. I have been with her for five years and I have never spent a birthday of hers with her. We have been married for a year-and-a-half and the morning of our first anniversary, I left for a month. So, as you can see she must love me.

Well I have to go. It is 2330 and I have a four hour guard shift starting at midnight. Thank you very much for the letter and I look forward to more in the future.

◆ ◆ ◆

The following letter is the first of two written by a major who served with the 2nd CER assigned to the Canadian Contingent of IFOR:

Somewhere in Bosnia-Herzegovina
February 6, 1996

Dear Mrs. Snailham,

Thank you for the letter, I was quite pleased to receive it. We are getting quite a bit of mail and the boys are extremely happy to respond. It would seem that entire schools are sending letters and trying to learn about this NATO mission.

It sound like you really enjoy Halifax. My family and I have been all over, but we have not yet lived in the eastern part of Canada. We have been posted all over Canada, but mostly out west and in Quebec. We spent three years in Germany with the NATO forces and we have lived in Cyprus for 12 months while I was the UN Force Engineer. I now find myself here in Bosnia for the third time — the first two were for very short periods.

You are right, Canadians are the best at what we do, for I have seen it first hand and I can compare it to the [response of the other participating] countries. We have a hard time selling ourselves to our fellow countrymen at times though. Maybe this mission will change that.

Life for the engineers in this mission is very busy and you are right, the threat of mines, boobytraps and the other instruments of a nasty war (in some cases, a civil war) are all around, but we have a job to protect our soldiers and we as sappers are proud to do what we can.

The infrastructure of this country has been badly damaged, and we are extremely busy trying to fix the power, water and roads. The locals are still happy to see us (unlike the UN mission where the kids started to throw rocks) and are helpful in telling us where the dangers might be.

At the moment, the weather is cold and snowy. The roads are slick and travel by vehicle is slow and dangerous. We are all looking forward to some warm weather and clear roads.

Well, I should end the note here. Thanks again for your letter; it was very much appreciated.

⬣   ⬣   ⬣

The following letter is the second of four written by a naval officer who was mentioned earlier on pages 51-52, and who commanded a frigate assigned to Operation Sharp Guard:

**HMCS *Fredericton*,
docked in Souda Bay, on the Greek island of Crete
February 9, 1996**

Dear Jane,

As you observed though, I have been exceptionally busy and have a hard time in keeping touch with my wife!

I believe that the last time I wrote, we were just about to go into Pireaus, Greece, early in January. That being the case, I'll update you from there.

Pireaus was very interesting. I took the train to Athens to see the Parthenon and the Acropolis, then went with some of my officers to view ancient ruins in Corinth, Akrocorinth, the Corinth Canal and other places with names too difficult to spell from there, we went to the city of Naples for a five-day visit. We hosted a

reception for the Canadian Embassy from Rome and met lots of high-ranking officers from various NATO countries. I think I had six admirals on board at one stage!

For the rest of the visit, I managed to get in two very poor rounds of golf (first time in almost two years) and went to see the ruins of the ancient city of Pompeii. That was a real education and very fascinating. On January 15th, we sailed to join up with eight other ships for three weeks of very intensive, coordinated multi-ship training exercises. I spent about 16 hours a day on the bridge and ended up eating most of my meals up there from the 18th until the 25th. On that day, the six ships of NATO's STANAVFORLANT of which [my ship is] part, pulled into Vallencia, Spain for a four-day visit. This was the first visit by a Canadian ship to Spain since the illustrious Turbot War [the writer is referring to a 1995 dispute between Canada and Spain over Spanish fishing fleets operating in Canada's territorial waters — editor]. As far as that goes, it was a non-event, but the city and people were great. It's also flat, so it was perfect for bicycling. The weather was great, about 17 degrees Celsius. On January 29th, we sailed again for another three-day exercise and then pulled into Polma Majorica for a few days. This was quite a hectic port visit, with a lot of organized social activities, so I never did get in to do any shopping.

From there, we sailed on February 5th, and did another two days of intensive training, then we detached from the squadron and hustled off to the "instep" of Italy where we sent our helicopter ashore to pick up some people who are joining the ship, plus about 4,000 pounds of stores, mail, parts, etc. After two trips, the helicopter was back and we headed off at high-speed towards Souda Bay for a 12-day Self-Maintenance Period. We arrived yesterday and took on fuel.

I have let about one-third of my crew return home for a mid-deployment family visit. I'm not going to be able to go, though. It just doesn't seem right to leave my ship in-theatre without me being there. I'm lucky I have a very supportive and understanding wife. Otherwise, I would be toast!

Well, my plans for the next 12 days are to finish up the huge backlog of paperwork I have, help the lads paint the ship and then wander about the island to see the sights. I hope at any rate, I will be able to get a decent nights' sleep.

Well Jane, once again, thank you so much for writing and understanding my somewhat spotty record of returning correspondence. Oh, as well, thank you for the "angel." Over the past few weeks, I think there has been a few occasions where somebody has been looking out for me!

◆　　◆　　◆

The following letter is the first of three written by a major who served with the Headquarters and Signals Squadron of the Canadian Contingent assigned to IFOR:

Camp Coralici, Bosnia-Herzegovina
February 9, 1996

Dear Jane,

I am a staff officer at the brigade headquarters, and have been putting in some rather long hours ensuring, not only as the Canadian contribution to IFOR arrives and starts its task, but also the other nations that form our multinational brigade. So far, the brigade has Czechs, Americans, British, Dutch and New Zealanders with more to come, no doubt.

We have established our headquarters in a small town north of Bihac called Coralici. The town and surrounding areas have survived the war fairly well and the local people seem pleased that we are here. The other Canadians are located in Velika Kladusa on the Bosnia-Croatia border and in Kljuc (about two-and-a-half hours south of Bihac). This is my first tour to the former Yugoslavia and I must admit it is a real eye-opener and nothing like I expected.

I expect to be here until mid-July 1996, when I will return to Petawawa and my wife and son. Our families are well looked after in Petawawa and it makes it easier to know they are safe. To date, the "factions" as we call them are complying with the terms of the peace agreement, but the real test will come in the spring and early summer.

Once again, thank you for your letter and please feel free to write again. As my schedule becomes more routine, I should be able to write more frequently. Take care and pray for peace.

◆　　◆　　◆

The following letters are two of five written by a 36-year-old master-corporal who served with the Headquarters and Signals Squadron of the Canadian Contingent assigned to IFOR:

Camp Coralici, Bosnia-Herzegovina
February 9, 1996

Dear Jane,

I must admit to being surprised at getting some mail from a "pen-pal." You shall be the first.

As you know, I'm in the beautiful country of Bosnia. We're stationed in a camp called Coralici (after the neighbouring village). The troops have a few nicknames for the place, among them being Camp Krusty (from the Simpson's TV show) and Stalag Coralici (stalag being prison in German). There is a few other nicknames for this place, however, those are the most polite.

The camp is surrounded with mountains which makes the place quite picturesque, but not exactly the most tactically sound. The biggest problem for the troops over here so far had been adjusting to the driving. The camp has seen some snow in the past few weeks and the roads are not exactly up to the Canadian standard. Once the boys learn to slow down and let the locals get around them, the incidents of vehicles in the ditch should (hopefully) come to an end.

I'm 36 and married, but don't look a day over 29, thanks to good clean living. I still have trouble figuring out how long we've been married as I did the dirty deed twice. I tell my wife when she's bugging me that my first wife never treated me the way she does. It takes a few minutes for it to click in, then the laughs.

I'm a bit of a movie/video buff back home and the hobby I like the best is home brewing. You get to enjoy the fruits of your labor. I haven't started making wine yet, however, I have got my two older sisters busy making it. It's fairly easy if you haven't got involved with the hobby and inexpensive after you buy the initial equipment.

I've been in the military since 1980. I actually joined up the day after April Fools. I'm in the radio operator trade which is basically the jack of all trades in the communication world. We work with a wide assortment of equipment. Actually, technology is slowly getting into our trade. We have computers, satellites, etc. It's a very full encompassing trade. Hard to get good, really good at anything because you change hats so frequently.

The days here are blending into one and about the only way other than looking at your watch to know which day it is, is church parade on Sundays. Considering the size of the camp, the turnout is greatest, probably, about 25 to 30 people which will no doubt get lower as the Protestant padre is finally in-theatre.

I would have to say that's about it for now. It's close to supper here and my date is coming to get me. One of the clerks is also married and we do the couple thing to keep the rest of the camp at bay, plus it's sociable.

——————————— ◆ ———————————

Camp Coralici, Bosnia-Herzegovina
March 5, 1996

Dear Jane,

I just got back from Budapest to find your letter in the office waiting for me. What a nice surprise it is coming back to the camp to some mail. Truly, a morale booster if there ever was one.

If you ever get the chance to travel over here for whatever reason, I would strongly suggest you check out the tour company. I'm sure the tour guide (from the tour company the military uses) was getting kickbacks from every restaurant and shopping concession we visited. Although I enjoyed myself, the fact my wife was still in Canada, and I had to figure out the gift buying by myself put me off a bit. The biggest reason for my apprehension was you never knew whether you were getting a genuine Hungarian article or if it was something made in the basement by slave labour.

I think the next time I go to Budapest, I will research the choice spots as we really didn't get the chance to spend anytime enjoying the historical sites we visited. I think I still have a copy of *Europe on $20 a Day*. I used that book to go to Munich with my mother when I was posted to Germany and the book truly gave you an idea of what and where to go and do.

I always hate talking about the weather because the last few times, the situation changed, I'm sure because I wrote home bragging about the spring weather. As I write this, the weather is sunny and borderline warm. In Budapest, however, it was far from warm and the wind whistled right through you. Some of the people were cursing their friends because they told them they didn't need a winter jacket.

One of the women on the trip is a clerk here in the unit, and unfortunately, she had a stomach flu throughout the weekend. She did manage to eat Friday and Saturday night, however, she didn't eat much. She did manage to get [out] and see the sights...

The trade I'm in basically involves passing message traffic from one point to another, either by teletype or radio. It's all a question of what part of the trade you are in. Some of the people in the trade [are] working with the satellites. The job I'm in involves controlling the access we have as a unit. Basically, vehicles, communication resources and people.

One of my friends over here does ham operator work back in Canada as a hobby. I'm interested in getting into the hobby on [my] return home. However, the guy in the unit who sometimes runs the courses tends to put people to sleep when he teaches. The last time I was Gagetown, NB, my buddy fired up his radio and was talking to someone in St. John's. I can't remember whether he got Halifax or not.

By the way Jane, thanks so much for the offer you made in reference to sending stuff over. About the only thing I'm interested in getting my hands on is a copy of

Europe on $20.00 a Day, so as to plan my next R&R (rest and relaxation) trip. A used copy of the book or even a photocopy of the section on Budapest would be great. Please don't kill yourself looking as I should be able to find something in April when I'm back for three weeks.

Bye for now Jane. I look forward to your next letter.

◆　　◆　　◆

The following letter is the second of two written by the major who was mentioned earlier on page 53, and who served with the 2nd CER assigned to the Canadian Contingent of IFOR:

Somewhere in Bosnia-Herzegovina
March 8, 1996

Dear Mrs. Snailham,

Thanks for the letter. Work here has just finished with the latest round of visitors. The Minister of National Defence David Collenette and the general of the army were here for a quick visit with the troops. I will be heading home soon as my work here comes to an end. A major will stay on to command one of my field squadrons and he would be happy to keep the correspondence going, with me going back. He will be the senior engineer in-theatre. That's all for now. Keep up the good work.

◆　　◆　　◆

The following letters are the third and fourth of five written by the 36-year-old master-corporal who was mentioned earlier on pages 55-57, and who served with the Headquarters and Signals Squadron of the Canadian Contingent assigned to IFOR:

Camp Coralici, Bosnia-Herzegovina
March 20, 1996

Dear Jane,

Thank you for the card and pin. I have all the bases covered now. I have the little angel on my T-shirt and I have the miraculous medal and the St. Michael medal on my dog-tags. Plus, I have an old prayer of protection in my wallet that my father-in-law gave me some time ago.

I must confess that the hours are indeed long, but the work isn't physically hard. The camp is pretty well set up now and the only thing left besides the normal work day is the on going beautification program. Basically, picking up candy wrappers and such. I didn't realize that our camp in Kljuc had made it onto television.

I would have thought that Coralici would have made [it] before Kljuc as this camp seems to attract all the big-wigs. We are pretty well the only camp that does roll out the red carpet for the visitors.

I can't believe that you went to the casino! And won on top of it. My wife was trying to convince me to go to [Las]Vegas, [Nevada, USA] on my three weeks. Could you imagine me coming off a seven-plus-hour flight and getting back on in a few days to go to Vegas? I told her maybe, after my final return in July.

I don't think I could have picked a better winter to come over here as my wife says, she hasn't seen much snow. She makes it sound like the only reason she wants me back is to shovel. I guess I know where I am in the pecking order.

One of the guys over here has the toonie [a Canadian two-dollar coin] or whatever they are calling it. Its just a matter of time, Jane, before most of the money goes to coins. Just look at the Germans. From pennies right up to $5, all in coin form. I don't understand the thinking behind going to coin from paper.

Changing the subject, I'm off to Zagreb tomorrow to take someone to the airport. One of the clerks here was a leftover from the UN days and is finally getting to go home. She's a nice person and all. I've been letting her aggravate me lately, because all she can talk about is her plans for her boyfriend on return to Canada. That's the last thing I need to know or hear about!

The Catholic priest over here is a friend of mine from Petawawa and he was suppose to come with us as he is looking for finger rosaries. Some of the people in his camp want some, and of course, there is no such thing in the military supply system. I told him we would be taking the clerk to the airport and that he would have to sit in the back of the truck with her. Most people who don't know him, can't believe he's a priest, good guy and quite a host.

I was told the other day that I'm finally leaving Petawawa. I was told that I've been in Petawawa long enough and it was time to move on. I'm hoping to go to BC; however, I've learned not to hold my breath.

I should warn you Jane, that I have a tendency to be a little bit of a devil's advocate. When I was born, my mom took one look at my eyes and knew she had her hands full. The only reason I mention it, is in the event that some of my sarcasm sneaks into the letters, you won't be shocked.

I'm enclosing one of the miraculous medals I mentioned earlier. My father-in-law had sent them over and most people don't put much faith in such things, so here you go. The medal has already been blessed. I hope you can find a spot on your person.

I'm going to sign off. If you see my wife say hello and tell her, I'll try to see her back in Petawawa in April sometime. Say hello to your son and husband and thanks again for the prayers and well wishes.

PS — Out of curiosity Jane, how do you remember what you wrote to who? I'm having a hard enough time keeping track, and I'm using a computer! Fill me in on the secrets of pen-palsmenship.

————————— ◆ —————————

Camp Coralici, Bosnia-Herzegovina
April 1, 1996

Dear Jane,

Thanks for the card and pin. You realize of course, this makes two angels on my shoulder. Did you want me to give one away to some deserving soul or put it on my other shoulder?

As you may be surprised to hear, I have been writing some articles for the military paper back home. No one else seems to want to bother and sometimes the ideas just flow off my fingers, so it's not a big problem. The part I dislike the most is the editorial scratches my work receives. You wouldn't want to write anything that points a finger at some of the stupid things that go on here.

My wife is starting to get antsy about my return, just another nine days and I'm home for three weeks. Don't worry yourself about the travel information as my mother is planning on getting something in Montreal for me. At least, that's the latest thing she mentioned on the phone when I last called. If worse comes to worse, I will look around the library back in Petawawa once my wife goes back to work.

On return to Canada in July, I will have approximately one month left and then we move to [CFB] Trenton, Ontario. The rumours of going east or west proved false. I [am] not exactly sure of the job requirements; however, it will give me more time with my wife. Hopefully, she will take some time to rest as she works midnights, four times a week at her current job.

Congratulations on the book. Will my wife and I get complimentary copies, signed by the author? I [am] surprised you've never mentioned the internet, Jane. With all the writing to UN personnel, I would think that a computer would come in handy. I realize that it might take away from the personal feeling of a handwritten letter; however, most operations now have access to an internet account. This would speed correspondence and also give you a change to gain a wealth of information you never expected out there.

I'm planning on getting an account, once I'm back home. There is too much going by to stand on the sidelines. Although I'm still unsure of my wife's plans for the summer and coming year, I think I can say with some certainty that we would be going east. The move to Trenton will prove enough of an adventure, I'm sure. I hope you're well and not up to your elbows in spring cleaning. Bye for now. Don't worry about me not being here for your letters as three weeks will prove to go by very fast. Happy Easter and thanks for the picture.

PS — I'll try to find a photo of myself that doesn't look like it belongs on a post office bulletin board. Ha! Ha!

◆   ◆   ◆

The following letter is the second of three written by the major who was mentioned earlier on page 55, and who served with the Headquarters and Signals Squadron of the Canadian Contingent assigned to IFOR:

Camp Coralici, Bosnia-Herzegovina
April 7, 1996
(Easter Sunday)
Dear Jane,

Things here have pretty well settled into a routine, but just when you think you can relax something comes up to change that. Right now, we are dealing with the removal of police checkpoints to allow civilian freedom of movement. The intent/ hope is this will allow them to return to their villages and try to get their lives back on track. Also, we are dealing with the last official IFOR mandated peace agreement timing — the cantonment and demobilization of the various faction armies. This is supposed to be complete by April 18, 1996 [D+120 or 120 days after the signing of the DPA], but will likely go on a bit beyond this time.

You asked what my job is. I am on the brigade headquarters staff responsible to the brigade commander for the current (or daily) operations of the brigade. The brigade is about 2,500 soldiers strong from seven different nations: Canada, the Czech Republic, Greece, New Zealand, the UK and the USA).

A few weekends ago, I had a chance to spend four days in Budapest to rest and "recharge my batteries." It was a nice break to eat some non-military food and wander the streets. I must admit it was difficult to leave and return to "war-torn" Bosnia. You can see signs of people trying to make the peace last, and I hope it works, but only time will tell. Pray for peace.

◆   ◆   ◆

The following letter is the last of three written by the major who was mentioned earlier on pages 49-50, and who was stationed at the headquarters of the Canadian Contingent assigned to the UNPF:

Tuzla, Bosnia-Herzegovina
May 1, 1996

Dear Jane,

I received your package and cards just recently. The first of your care packages arrived on March 15th, just as I was going on leave to Italy. I was in Zagreb awaiting a flight to Rome to meet my wife. One of the people at the Canadian

Contingent office in Zagreb gave me your package and one from another friend at the same time. The package from my friend was fudge that they mailed in late December. I was afraid it would have festered and caused an outbreak of a new disease that would have endangered the world. Their gift made me very popular with the people in the Canadian office in Zagreb, in particular, a young refugee lady. She is our cleaner and very highly educated, but as she's from Sarajevo, the Croatian authorities don't recognize her qualifications.

Your gift, however, I was very selfish about yours. I stayed overnight in Zagreb and flew to Rome on March 15th. (I was standing in front of the statue of Julius Caesar in downtown Rome on the Ides of March). I didn't let anyone have any jujubes (my favorite!) and I killed a few hours with your crossword puzzle book. I am a crossword addict, and I have been going through withdrawal very poorly. Thank you so much for your kindness. You're a real sweetheart.

Tuzla - an interesting location! This is the headquarters for the US division (the 1st Armoured Division). Under the DPA, Bosnia had been divided into three sectors, the upper right hand section is Multinational Division (MND) North, and is administered by the Americans, the lower section, Sarajevo and south is French, (MND South) and the western section (MND Southwest) is British. There are 32 nations participating in this operation and we have 11 in this sector.

The main focus of this operation is to staff the ZOS. [The] Inter-Entity Boundary Line (IEBL) is the line that divides the country into Serbian and Muslim sector. The Serbians are supposed to have 49 per cent of the nation and the Muslim and Croats are supposed to have 51 per cent. The peace IFOR came into Bosnia and solidified the IEBL by putting up checkpoints and establishing a zone two kilometres wide on each side of the IEBL. This four kilometre wide zone is patrolled, and anyone passing through the line by any of the roads must go through the IFOR checkpoints. If someone is carrying a weapon, it is confiscated. If large groups try to go through the IEBL to cause trouble, then IFOR won't let them through or will, but with provisions to restrain their activities.

The ZOS is really the battle line at the end of the war and is loaded with mines. There are estimates that there are at least three-and-a-half to six million mines in Bosnia. The impact of all this is that everyone has to be careful walking anywhere, particularly in the ZOS. Mine clearing is the responsibility of the factions, or the "entities" as they are known in the DPA. IFOR supervises the clearance of mines. You can't really go into a minefield that someone else has laid and clear it and expect to come out with all your limbs intact. The entities have laid the mines and have mapped where they are, in most cases. Some have been boobytrapped, some have anti-handling devices and some have been laid by weirdos who put them in the ground at a slant so that when you prod for them you trip them, and then it's "goodnight, Irene." So, we have the people who laid the minefields go in and clear them. However, many of the Bosnian and Serbian soldiers here were given extra anti-personnel mines, many that look like green hockey pucks. No one touches these, they are blown in place with plastic explosives. So they would lay these mines to protect their positions. These aren't mapped and even the people who placed them don't know where they are anymore.

I was visiting Coralici where the Canadian Brigade is located. Minister of National Defence David Collenette was visiting the forces and I joined the other two Canadian PAO's to be on hand when he visited. We were at a mass grave site where they found 11 bodies. I was standing by the side of the road and I was absentmindedly moving a round rock that I was standing on.

I very quickly realized that this was too round to be a rock and I looked at it. It was a bomblet that hadn't exploded. I've had several situations like that and I'm careful. So you can imagine how many people who aren't aware of the danger of mines are going to be killed by them before the minefields are neutralized.

We've had some problems here during the past few days. Today is the last day of the Bajram (pronounced By-ram), a holy period [lasting] days. Under the guise of visiting their former homes and cemeteries, large groups of Muslims have been crossing the IEBL and their movement has been opposed by large numbers of Serbs. I'm talking about groups of several hundred people at a time. At one of these "demonstrations," a woman was pushed to the ground and a Serbian man was about to behead her with an axe when a Danish soldier fired a flare pistol at him and struck him in the head with the cartridge. He was probably given a headache for several days, but both are still alive.

I don't understand the hatreds that are here. This is a country that has animosities so deep that violence is an accepted form of political activity. We all have to be very careful when we go out.

I don't feel any threat when I go to Tuzla, but if I go beyond and toward the ZOS, I always take my flak vest and helmet. It's an interesting experience and very different than when I was here with the UN last year.

My current boss, a Norwegian army officer, wants me to extend for the full time of the mission, but I told him that would keep me away from home for 18 months, and that's simply too long. My wife is extremely patient and always supports everything I do, but I want to get back to my family soonest.

I'm using [the] internet to keep up on Canadian news. I only discovered that we had [the] internet yesterday and immediately got onto it.

I've got a few things I have to do, so I'd better run for now. I hope that the outgoing mail is quicker than incoming, cuz it takes about two to three months for me to get mail. This is what I call "snail mail."

Cheers for now, and thanks again for all your care. I remember you in my prayers and I still wear your guardian angel on my right shoulder.

◆ ◆ ◆

The following letter was written by a sergeant who served aboard a frigate assigned to Operation Sharp Guard:

HMCS *Halifax*,
on patrol somewhere in the Adriatic
May 8, 1996

Dear Jane,

Let me first begin by telling you a little about myself. I am married and have two adorable children, a son and a daughter. My wife and I, are from Cape Breton, where we met in high school. We were married in September 1982, following her graduation

from nursing school. I joined the military in March 1979 and I am currently serving in my 18th year. I began my career in CFB Shearwater and have served all of my time here, with the exception of a short three-year posting to the Aircraft Maintenance Detachment Unit [at CFB] Trenton in Ontario.

In your letter, you asked me to explain what my job is, well, here goes. Our [unit] consists of 19 personnel, eight of which are aircrew, four are pilots and four are navigators, with the remaining known as aircraft maintainers (technicians). The top dog is called the Air Detachment Chief, that's my boss. I am his second-in-command, which leaves nine other technicians of various aircraft trades, who work directly for me. It is my job and responsibility to make certain that the aircraft is properly maintained and ready for flight at any and all times. As we began this trip, it seemed like a long time to spend away from our families. Throughout this deployment, all personnel will be given an opportunity to return home for a short period, I will be home while the ship is in port in Roosevelt Roads, Puerto Rico. After my visit home, there will be only six weeks remaining until we sail into Halifax, NS.

Once again, I thank you for your time in writing to me while on deployment in the Adriatic, off the coast of the former Yugoslavia. Take care.

◆   ◆   ◆

The following letter is the last of three written by the major who was mentioned earlier on page 55 and pages 61, and who served with the Headquarters and Signals Squadron of the Canadian Contingent assigned to IFOR:

Camp Coralici, Bosnia-Herzegovina
June 9, 1996

Dear Jane,

I was away in Canada on three weeks leave. As expected, it was great going home, but very difficult to return here to Bosnia. While in Canada, my wife and I spent three days in Ottawa at a bed and breakfast, and two days (including Mother's Day) at a very lovely country inn near Sherbroke, Quebec. We were supposed to have stayed longer at the inn, but I started missing my son, so we cut our stay short and returned to Petawawa. My parents came up from the states to look after him while my wife and I were getting "re-acquainted." Everyone survived the ordeal and my parents said they would do it again. The remaining two weeks were spent with my son and doing the odd jobs around the house.

I am due to leave Bosnia, for good, on July 10th. We are being replaced by soldiers from Valcartier, Quebec. They will have a difficult time I suspect because it will become more and more difficult as time goes on. The military aspects of the peace agreement have basically been achieved, but implementing the civilian aspects will start to take on more and more importance and may be the cause of the peace

agreement failing. For us, it was easy because dealing with soldiers, no matter how professional, is relatively simple. But getting civilian organizations and governments to agree is much more difficult and may lead to problems down the road. Especially when IFOR pulls out after a year.

When I get back to Canada, I am entitled to a few weeks off, which will be nice. We have not made big plans, but [I] look forward to getting back to being a family again. So much can happen in six months. It will take some time to get back to normal.

You mentioned Canadian soldiers helping the locals in your letter. Right now, we are helping to repair roads damaged during the war, clean up the garbage from "their" war, teach about the dangers of mines to the local schools and in a limited way distribute aid to the areas that need it the most. This all seems to be helping as there are signs of people and their lives getting back to normal all around us. Let us hope they become accustomed to this way of life and stay at peace. But only time will tell.

The weather here is very nice. Right now [it is] 20 to 30 degrees Celsius which makes for a long summer with sunny weather. Something Canada, especially Petawawa, could use. Today is Sunday and we always try and have a bit of a "day off." It gives me a chance to write a few letters and do some things for myself.

Well take care, and thank you for your letters. I appreciate you taking the time to express your concern and gratitude. Pray for peace.

◆　◆　◆

The following letter is the last of the five written by the 36-year-old master-corporal who was mentioned earlier on pages 55-57 and pages 58-61, and who served with the Headquarters and Signals Squadron of the Canadian Contingent assigned to IFOR:

Camp Coralici, Bosnia-Herzegovina
June 21, 1996

Dear Jane,

What a pleasant surprise to get another letter before heading back to Petawawa. I think I have about 11 days and a "wakie" left to go. I still don't know who exactly is coming over that I might know, however, some of the 5th Brigade are bilingual, so maybe you will get lucky with one of them, pen-pal wise.

Things are starting to wind down for us with the anticipation of the new blood coming in, however, most of the people are slowly driving themselves crazy thinking about how much time they have left over here.

Naturally, we have parades coming up to eat up some of the time, so the last few days for me will pass quickly. There is supposed to be a medals parade either on the 29th or 30th [of June]. Then we have a change of command parade on the second. Lucky me, I'm invited to both!

Thanks for the well wishes in regards to the house. If only, my dear wife would calm down a bit and try to enjoy the fact the work is all done, then everything with her would be fine. I think the problem is that too many of her workmates are pestering her with the same questions — Is the house sold? When is your husband [coming] back? Did you find a place in Trenton yet? It goes on and on. I foresee much work on my part trying to put her back into her old frame of mind.

It's amazing how fast the time seems to have gone. I know the time didn't go too fast for my wife, however, I don't think she's any worse for the wear. I know she wouldn't be impressed if I volunteered for another tour in the next year or so. If I have my choice of going somewhere again, it would have to be the Golan [Heights between Israel and Syria], as the CO was telling me about the fabulous tours of the Hold Land available there. Tours that would cost several thousand dollars if you took them from Canada. Something tells me that my wife wouldn't be too pleased to find out that I have volunteered to go on another tour just for a free trip to the Holy Land.

Well Jane, thanks again for your support throughout the tour. It was great for my morale getting the letters and well wishes. Rest assured that I will keep you posted on my return to Canada.

⬡　⬡　⬡

The following letter was written by a major who served with the Media Operations division at the headquarters of the Allied Command Europe Rapid Reaction Corps assigned to IFOR:

Sarajevo, Bosnia-Herzegovina
August 12, 1996

Dear Jane,

Having just come back from my first 90-hour pass after two months in-theatre, I figure this is as good a time as any to put pen to paper, so to speak, and provide an update on the goings on in this part of the world. Now, you may note that this appears to be a photocopy, which it is, and for that I apologize, but in the interests of trying to keep family and friends moderately happy about the flow of information from this end, I've had to resort to these means, lest I spend every waking moment of my life in front of the computer writing letters. I'm hoping you understand and still write back, if so inclined!

Impossible to believe, but the day after tomorrow will mark my eighth week in-theatre; it seems inconceivable to me that two months of my life has passed so quickly, but obviously there have been a lot of experiences on a daily basis, both fulfilling and frustrating, which makes time fly by, so let me start filling you in on some of them. Figure the best way to start is to describe where it is I am, put that in context and share with you my daily routine.

I work in what is known as the corps headquarters, or ARRC (that stands for Allied Command Europe Rapid Reaction Corps) HQ to be more specific; which is the main headquarters for the ground forces in IFOR. There are about 53,000 soldiers here, which is kind of like taking all of Canada's army, multiplying it by three and putting that into Ontario. The HQ is based in Ilidza, which is a suburb about 10 kilometres west of Sarajevo. The rest of the troops are spread throughout Bosnia in three divisional areas — to the southeast (HQ: Mostar) are the French, to the southwest (HQ: Banja Luka) are the Brits (and Canadians) and to the north (HQ: Tuzla) are the Americans. There are 34 different countries participating, which can make things interesting to say the least. More about that later! Anyway, in this HQ complex there are about 1,400 soldiers, 99 per cent male, so there is a lot of testosterone to be sure, especially with all these special forces and special forces wanna-be's around! I work in the media operations section with a Brit colonel, a French lieutenant-colonel, an American civvie media advisor, two clerks from Glasgow and two Brit drivers. Our office is the "nerve center" of the media operations since we service the commander of the ground forces (British Lieutenant-General Sir Michael Walker) and act as the primary spokespersons for the whole mission. Over the course of the day, we do briefings to [the] media and the commander, write responses to queries, arrange interviews for journalists, track down information, plan strategy and read voraciously.

Our workday runs from 0700-midnight, seven days a week, though there is no press briefing on Sunday and it tends to be a little calmer in any case since most of the journalists take off that day. That being said, tomorrow (Sunday), the Deputy of the Office of the High Representative, is giving a media brief at 1100, so we'll be there for sure. Here's the routine:

[I am] in the office around 0700 or shortly thereafter, where I review the operational report, the situation report and the intelligence brief, all of which review operations conducted over the course of the past day.   These give me the basis to prepare the script for the press conference. At 0800, about 50 [personnel] gather in the commander's conference room to get an update from all functions and divisions about what happened over the past 12 hours. I then wildly scramble from desk-to-desk trying to extract more information from sections for use in the brief. From 0830 to 0850 or so, our section goes to the Command Information Communications Group meeting which is our chance to sit down with the General and select others to review media items of interest and to get the commander's feel on how he wants to place certain issues; this provides valuable feedback, since it comes right from the horse's mouth and is a clear indication of how important media relations are to the operation. With this I can finish up the brief. At 1020, we head out of here for the Holiday Inn in downtown Sarajevo, where we meet with the other various spokespersons (from the High Representative, UN High Commissioner for Refugees, UN Police Task Force, Office for Security and Cooperation in Europe, etc.) and talk about what it is we are going to say at the news conference, with an aim towards ironing out any inconsistencies in our stories before going public. At 1100, the show begins to about 40 journalists representing just about everybody who's anybody — *CNN, NBC, Reuters, Agence France Press, BBC, Newsweek, the New York Times, the Washington Post*, etc. Unfortunately, the *CBC* [Canadian Broadcasting Corporation] is a very bit player in this field, and one of the reporters for *NBC Radio* does some freelancing for *CBC Radio* whenever there is a Canadian scandal to report on, and another does some freelancing for [the] Canadian press, though reports from here for Canadian audiences, I can count on two fingers since I've been here.

We do six news conferences a week, and the Brit lieutenant-colonel and I alternate as spokespersons for IFOR. I do Monday, Wednesday and Thursday and he does Tuesday, Friday and Saturday. He's away on leave right now, so I'm doing them all for two weeks straight. I've done about 10 of them so far, and am becoming more comfortable about doing them, but the same can't be said for the first couple! Despite the fact that we do this full-time, we don't have a lot of practical experience doing interviews on a regular (i.e. daily) basis, and I can assure you that doing an over-the-phone interview is a far cry from doing something in front of CNN cameras! There are also about 75 other people at the briefs, from all sorts of NGOs [non-governmental organizations], to embassy folks, to senior military, since that is "the" spot to get the latest goods on the political and military scene as it develops. The transcripts of the remarks is also put on the internet that night so the whole world can see what an idiot you've made of yourself! So far, I've escaped unscathed, and have done okay; you certainly don't learn all there is to know about an organization this big, in a situation this complex, in a few short weeks, but I'm giving it the good old college try and just do the best I can. That wasn't the case for my first day out, for sure; good thing the transcript reads better than the carnage that day during the session! Needless to say, I was just a little apprehensive, with but four weeks in country and expecting to represent the commander on all things

IFOR-related. I read the prepared text (well, I think), and then came the dreaded questions. The first one was about the general situation with respect to freedom of movement, and the IPTF [International Police Task Force] guy and the High Rep were asked the same questions, and then I was as well. [For example,] "How come IFOR isn't doing more to help?"

The subject was a little stale by then, and I said that these things take time, yadda, yadda, yadda, which seemed like a perfectly reasonable answer and was already patting myself on the back for, when about seven hands shot up and one of the hands asked me, "So is it the Pentagon's position that this is a civil war?" This seemed so stupid a question as to be dangerous! I said I couldn't speak for the Pentagon (no good, that lasted about one-eight of a second before it was, "Okay what does IFOR call it?", so I said well, a civil war in the sense that no outside countries invaded a single country, which then become several countries, ergo a civil war — as it turns out, we don't have a position, since the Muslims call it a war of aggression and NATO just calls it a conflict. So, it went on like that for about a half hour, after which I figured I probably was going to be the shortest-lived NATO spokesperson in the mission! But the journalists cut you a bit of slack initially, since they realize you are brand new, (some of them have been here for four years covering the same story!), and, by then I had the opportunity to get some info for them and chat over coffee, so I had begun to cultivate a good relationship with them — they weren't about to burn me on the first day! I suppose it wasn't that bad, though it sure seemed like it at the time. All I can say [is] I'm getting better at it, but have never been one for the spotlight, so can't say I particularly enjoy the parry and thrust in that forum. I prefer developing strategy and devising the lines to use for someone else to deliver. Have to admit, though, it is kind of neat when you come up with a good answer to a tough question, and see that reported in the world's media. Now, notice I didn't say Canadian media, who are really quite pathetic in comparison to the world standard. They are extremely immature, in outlook and subtlety; geez, they couldn't even be bothered to come here, instead choosing to take their copy about Bosnia from the wire services, which is much cheaper for them, to be sure. Seems a bit criminal, in a sense, given that there are 1,000 of our guys doing a bang up job in the northwest part of the country...but just wait for it if anything goes wrong.

On that note, don't expect to see me too much on Canadian news, since I am a spokesperson for the whole shebang here, not for the Canadians — they are but one of 34 nations, and a "bit player" at that, and anything to do with them is left for the Canadian PAO as it is small potatoes for me here. We get involved in the macro-events, the whole issue of NATO policies and the operations in-theatre, though, I have done a couple of pieces for *CBC Radio* when they couldn't get a hold of anyone in the Canadian sector and wanted a Canadian to talk about what had happened, for example, the fire we helped put out and the guy who died in the traffic accident. My job is sometimes described as trying to turn chicken shit into chicken salad! After the news conference, its back to ARRC HQ for lunch, and the rest of the afternoon is spent reading reports of every available description and attending to the daily routine of media queries, arranging interviews, etc. At 1900, we have the commanders call, which is the same as the morning call on a wider scale. That normally takes about 40 minutes, then it's off to supper, and then back to the office to do more reading, assessing, etc. till about midnight. Within all that are other meetings of various sorts, including [one on] perceptions (we look at what perceptions we are creating in our target audiences, and what it is we want to create); the Information Campaign Matrix (I collate what other people are doing for a master list of products and events); the Joint Information Campaign Coordination Conference; and a host of other minor bits.   We also do a weekly slot with the Republika Srpska [the Bosnian Serb Republic] radio and television station, and others throughout the week on an as-required basis, in addition to keeping abreast of the latest operational situation and providing input into that during the planning stages. This is very interesting, since you see how a mission begins right from the start, some of which have the potential for world-news making effects! So you see, it is a rather complete day and night.

To try and explain what is going on here would take a book, or two or three, so I won't bore you with the detail. Suffice to say that this place is seriously screwed up. Just when you think you've got something figured out, another twist comes by which throws you completely for a loop. So, for example, one of our patrols gets a pot-shot taken at it in the wee hours of the morning the other day, near the Visoko airport, which is in Muslim-controlled territory. We move into the area the next day, find six guys, two in uniform, with lots of weapons; of course, we took everything into custody. Looked like a great news story — we turn them over to the local authorities for [pressing] charges. Well, it turns out the mayor of the town, to whom the police chief reports, is also the airport manager, and it is him who has established this protective posture in the first place. Good luck seeing anything happening there. Or, each side uses the humanitarian aid they get to pressure people into voting in particular places; for the Muslims, they want their people to vote in the community they were [in] before the war, and in so doing, hopefully elect Muslim mayors and town councillors in areas that are now under the control of the Serbs, whereas the Serbs want everyone to [vote] where it is they are now, so are trucking people up to key locations in the country and dumping them there to live (taking over Muslims apartments, of course) and thereby stacking the vote in place they really want to win. And hey, if you don't want to play, they don't give you aid. Tends to be a real arm-twister when the male breadwinner has been killed and you're stuck with two or three kids to feed and no place to call home anymore.

Interestingly enough, the wife of Radovan Karadzic also happens to be the head of the Bosnian Serb Red Cross (his daughter, by the way, is the chief of information). All pretty cozy, isn't it. Everyday there is a funny story and a sad story, which makes it fun to be here. Awhile ago, for example, a mentally deranged Serb attacked one of our armoured personnel carriers with an axe, and the Serbs launched a protest against us! The only thing we could figure out was they were mad his axe was dulled or something. His name, as it turns out, is actually Bozo (pronounced Boshoe).

Or, today, a French corporal stopped by the river to jump in and refresh himself in the 40 degrees Celsius heat, hit a rock, and is now paralyzed from the neck down; the other day a Moroccan had his hand ripped off in a road traffic accident. This stuff does not even compare to the reports we read about what went on during the war. You just turn it off and become desensitized to it. Problem is, no one side is less guilty — the Serbs do some pretty stupid things to be sure, but the Muslims and Croats have been very good at propaganda, far better than the Serbs, so they appear to the West to be the ones [who suffered] the most. In fact, the Croats are probably the best at ethnic cleansing, the Muslims the most duplicitous of the three groups. Truth be told, they all deserve each other.

Okay, rather than bore you any further, here's some snippet of life and general observations about the place to let you in on what it is really like around here.

Weather: About 34 degrees Celsius everyday and really cool at night, though still warm enough for hordes of humongous mosquitos to do their damage during the evening. Actually, there's not that many of them, but they bite good — the clerk has a bandage on his thumb because he got so many bites on it, and they seem to especially like the forehead of my boss.

Accommodations: Ah, a little spartan. [The] first three weeks, I was in the transient quarters, waiting for the guy I was replacing to leave. Had seven roommates most nights, 11, if you count the cockroaches. One small bathroom. Thing is, the roommates would all change each night, so I think I was up to sleeping with nine different nationalities before moving to my new digs. Am in a room about the size of a medium bedroom back home, with two roomies, a Brit major (the weatherman) and an American sergeant. We used to have a fourth roomie, but he's moved to Tuzla for a bit, and we put an extra sleeping bag on his cot, so the room police will think there are four in there and therefore, won't put another person in. The Yank is in bed at 2100, and up at 0530, so there isn't much opportunity to do anything in the room even if you wanted to. End up using it to crash and store kit, that's all. I use two of my barrack boxes as a little coffee table for my clock, clothes (there are neither closets nor shelves) and Walkman. [I] sleep in or on my sleeping bag, and for a pillow [I] have wrapped my American poncho liner around my combat coat.

Laundry: Tried the system here once. Put about three-quarters of what I had in, and waited, and waited and waited. Eleven days later, got it back. Had visions of having to get my wife or mom to buy me underwear and socks and ship them over priority post! During that interminable wait, the others in the office got their's back smelling like it had been washed in sewer water. So now, once a week I give my stuff to the cleaning lady who washes, irons and folds it for 10 Deutschemarks (about $9 CDN). It's worth the piece of mind.

Clothes: Real easy. Combats and nothing but. Went on my first 90-hour pass the past week and wore civilian clothes for the first time in seven weeks. Really! Get up, put them on at about 0700, take them off about 2400 or so. Makes it real easy to pick what one is going to wear. Have been wearing the new tan Canadian boots since day one. They are the cat's bottom. Break them in, in 10 minutes, fit like slippers, have "fast laces" which means you can tie them in about 10 seconds. Our combats are comfortable, but do look a bit ratty compared to other countries, who are required to iron theirs. The office describes it as "combat tramp 65."

Food: Too good. The Brits run the kitchen and it is excellent, though they apparently have not heard of greens, milk or napkins. Getting a little tired of lemon and orange squash drink at this point. Have given up coffee, at last in camp. I drink cappuccino if [I am] out, (since the Turkish coffee is like liquid caffeine), since I am convinced the grounds are volcano dust or ground up brown crayons. Have put on a little weight, not too much, since [the] food is good and not much opportunity for exercise.

Water: The reason for increase in weight. Water is sporadic. Comes on around 0710 most times, which is too late to get any before being in the office, and may or may not be on throughout the day, and certainly off after 1900. Couple of times have gone out running in the afternoon, and returned to find the water off completely, without even a trickle of cold! Now, I don't care how much you abuse me, just let me have my hot shower in the morning and I can take the rest of the day. The fact of hot water on demand was probably the best amenity part of my leave! Well that, and the clean crisp sheets on a double bed with a real pillow.

Leave: Went to Dubrovnik, the most incredible city I've seen. Took the military flight to Split, stayed there overnight at a pensione [a small hotel] — there were six Lithuanian police officers from the IPTF there as well, so we drank some cherry brandy and chatted and watched some of the Olympics together. Took the bus to Dubrovnik the next morning, a brilliant ride along the coast with mountains on the left and the sheer cliffs to the right which lead to beaches and cute summer homes. Found a pensione which overlooked the old city, big balcony, five minutes from the beach and a landlady who kept plying me with cakes and coffee, all for $30 a night. Spent two days there, walking around in the morning, having a snack for lunch, coming back to get changed and then off to the beach for the afternoon. Lazed around there, diving off the rocks which form the base of the old city. [The] water is crystal clear and just cool. They [the residents] think it's cold — obviously [they] haven't tried the south shore! It is so salty, you float easier in the water, believe it or not. So, I would jump in and swim for a bit, let the sun dry me out, then swim, dry, etc. Back to the place for a shower, then down to the old city again to walk around and have a great seafood supper. Walked around the wall to the city, which is so huge it takes about 90 minutes to walk around. Some parts of the city are 900 years old, with a lot of it being at least 400 years old, absolutely spectacular. Then spent the night sitting in the cafe sipping real coffee and just watching the world pass by. Took the bus back to Split on Saturday, the city bus out to the airport, and made it with 10 minutes to spare, to get changed into my uniform and on the flight back to Sarajevo. While on the bus, somebody's fish crate leaked all over my rucksack, which reeked to high heaven — must have been quite the sight, me washing an army rucksack off in the main city fountain! Anyway, have to admit that it took a couple of days to get back into the swing of things here — the three-week leave we get will be interesting, as I can only imagine how difficult it will be to come back after living in the lap of luxury for awhile.

Multicultural Dimension: It is sometimes difficult to feel much for any of the former warring factions, given their track record and given their maneouverings behind the scenes to further their own cause. Now that is to be expected, but if I was a Serb, I think I would quickly come to the conclusion that there is a western conspiracy about — international pressure is almost solely against the Serbs to comply with Dayton, notwithstanding the substantial noncompliance by the Muslim and Croats (especially in the former, in the area of freedom of movement); the Train and Equip programme, which aims to send hundreds of millions of dollars of arms into Moslem hands, will tilt the military balance in their favour and by far the International War Crimes Tribunal has indicted far more Serbs for war crimes than Muslims or Croats, and don't appear to be too interested in following up on cases that are not Serbs. I am no Serb-lover, but they do appear to be getting the short end of the stick. The fact is that they are no military power, and Dayton saved their butt, but good; another week or two and the whole of Western Bosnia would have fallen to the Croats and Muslims. To equip the latter two with modern equipment should scare the Serbs very much. There are unbelievable and incredible violations of civil and human rights here; the police are so unbelievably inept and corrupt. Well, most of them are former military, who are recruited for their particular skills in fighting and torture, to be sure. Okay, enough of that.

Jane, I really appreciate your letters — you are an incredible person to sit and write to so many people in hand like that. If only we had more people in the country like you, we wouldn't have to read about all the crap in the news. It really is heart-wrenching to see what can happen to a country that lets itself get consumed with truly trivial and unimportant things.   I read most of the defence-related news articles from Canada, and am constantly sickened by what I am reading about the organization. I guess the thing that makes me (and others, I'm sure) so mad is that there is all this talk about leadership and doing your duty, but it doesn't seem to start from the top. Where were the generals standing up to the politicians when they put a personnel cap on our forces, which led directly to Somalia, and to potential problems here and in Bosnia; the politicians are sacrificing the security and safety of the Canadian soldier by this and other means. Though, I truly agree with your comments about how professional and good the CF are on these missions, I think there is a serious leadership problem, but it's not at this rank level, it's at the general level. Take a look at the accountability regime of the Americans where a number of generals get fired for not getting involved enough in their forces — good luck at seeing that here.

In any event, I am starting to ramble on here, and think I'm going to get some extra rest tonight, so, once again, thank you very much for your wonderful letters. I will try to write again as soon as I can. Once again, thank you for your time and thoughts. Cheers!

●   ●   ●

The following letter is the first of two written by an artillery captain who served with the LdSH(RC) assigned to the Canadian Contingent of IFOR:

Camp Holopina,
Coralici, Bosnia-Herzegovina
August 26, 1996

Dear Jane Snailham,

Greetings from Bosnia! My superior received your letter and decided to pass it on for all members of the Liaison Officer Cell to have an opportunity to write, if they wished. I thought I would write so that at least one person would respond to your kind letter.

I belong to the 1st Royal Canadian Horse Artillery Regiment, and [I] am an artillery officer by trade. I live in [CFB] Shilo, Manitoba with my wife of four years and my son. We have a three-bedroom, private married quarters (PMQ) on the base.

I was promoted to captain in April of this year, shortly after I received the opportunity to deploy to Bosnia. I work as the liaison operations officer for the Battle Group. There are approximately 900 soldiers in the Battle Group spread over an area about the size of PEI. There are three main camps: Holopina (located in Coralici), Maple Leaf (located in Zgon) and Drvar (located in Drvar). Camp Coralici was renamed to Camp Holopina to pay respects to a Canadian engineer NCO [non-commissioned officer]who was killed in Bosnia in the early years.

My job entails the coordination between other liaison officers in the Battle Group and the Battle Group Headquarters here in Camp Holopina. I conduct the daily administration for the Liaison Cell as well as coordinating the civilian interpreters daily. My day begins at 0730 and goes sometimes until 2200 (minus lunch and supper, of course) . I have to put out daily reports, which can take several hours to prepare.

Back home in Shilo, I play lots of golf (my favorite sport), and spend time with my family. I come from a military family, who moved from coast to coast. I joined the army in 1990 and went to Royal Roads Military College in Victoria, BC. I studied political science and economics while I was there.

My deployment here in Bosnia will last until mid-January. I have been here for almost two months now. I am going home to Shilo tomorrow for my three weeks leave.   That should give this letter time to reach you, and probably for you to respond in return. Look forward to hearing from you.

◆ ◆ ◆

The following letter was written by a major who served as the Deputy Commanding Officer of the 2nd Battalion, PPCLI assigned to the Canadian Contingent now assigned to SFOR:

Camp Holopina,
Coralici, Bosnia-Herzegovina
January 18, 1997

Dear Jane,

Well hello from not so sunny Bosnia! First thing I have to tell you is what a pleasant surprise it was to hear from a "stranger" (henceforth to be referred to as my "pen-friend") from back home. It really is a great gesture on your part to go out of your way to support those of us in Bosnia. And thanks, it really is appreciated. I promise I'll keep up my part of the deal. Hey, you've got it in writing now. I can't let you down!

I hope many of my compatriots are able to receive a similar offer of a pen friend — it's great to know someone else, other than your family, cares for you. But how did you come to pick me? I have a funny feeling my name was in the media somewhere and I'll get into how that might have happened in a bit.

Since you asked, let me tell you about my family clan. I have a lovely, and so far, very supportive, wife and two beautiful children. My wife and I were DINKs (double income, no kids) for the longest time until she woke up one morning and said, "My biological clock is ticking, I've changed my mind about being completely career-orientated." And the rest is history.

Being in the army, they are somewhat used to dad being away for a few weeks at a time now and then, but six months with only one trip home is something else! There's no doubt about it — it's not easy and we give our spouses fill credit for their support. The "soldiers" will get a medal for their tour here, but what about the spouses back home? Nothing but our respect and admiration for having put up with the absence. And you know that character Murphy, Murphy's Law — "If anything can go wrong, it often will at the most inopportune time." And sure as shooting, that Murphy character jumped in right after I had left. Although we had lived in our present house at [CFB] Winnipeg for a year and a half without a problem, I wasn't gone 10 days before the sink backed up in the basement and one week later my wife had a fender-bender (no injuries, thank goodness and minimal damage). It always seem to be that the fridge or stove or car breaks down when we're away on exercise or whatever. Can you imagine how helpless it makes you feel when you're thousands of miles away? Wait till I get my hands on Murphy's throat!

For the men and women here, it is probably easier on us than it is on our spouses back home. We have camaraderie, our meals are cooked for us and we work such long hours, the time literally flies by...while our spouses have to raise the kids, in some cases, work all day then come home and somehow find the energy to cook supper and look after the kids' emotional needs. Yup, no doubt about it — it takes some special kind of person to be married to a soldier (sailor or airman — men and women).

Now don't get me wrong, Jane, we wouldn't trade this mission for anything in the world. It's exactly the kind of fulfilling, personally rewarding mission we dream of. To have a chance at contributing to peace whether it's Bosnia or Timbuktu is extremely gratifying. Over here, it's incredibly satisfying to see the waves from the people as you drive by. We "live" for this kind of operation. Near our camp, we also have a chance to go for local runs to stay fit, and this takes us through many hamlets of five or six houses. During these runs, we go in pairs as a minimum (the "buddy" system — it isn't quite as safe as Canada yet) and the kids always run up to us as we go by to give us the high five. And this is just the tip of the iceberg. You'll never see on the news the kind of personal stories Canadians would be proud of, although I'm going to try to change that — I'll explain that in a minute. It seems the media would rather attempt to destroy our reputation bases on the conduct of a couple of "yahoos." I can assure you, we are doing our best to eradicate this kind of behaviour. The Canadian service personnel continue to be among the finest in the world, bar none, and you'll hear this from Americans, Brits, French, etcetera. But anyway, I ramble.

And since you asked to learn more about who we are — on January 10, 1997 we took over from our predecessor, 5th Brigade staff from [CFB] Valcartier, Quebec. We are known as the Second Battalion Group, PPCLI. We are based in Winnipeg, but enjoy significant augmentation from all across Canada, although about 98 per cent of our augmentees are from Western Canada. There's 900 of us here, but this does not include two other groups of Canadians — the NSE of 250 personnel, and the National Command Element of about 20. If you'd really like to learn more, I'd invite you to get on the Internet by whatever means you can and go looking for the "2nd PPCLI." As I write this letter, we're about two days away from having it up and running, so unless there's some unforeseen obstacle, it should be there by the time you receive this letter.

My job here is as the second in command of the Battalion Group (we call it DCO-Deputy Commanding Officer) and some of my responsibilities include working with the media. Now you understand some of my earlier comments! There is no one who has more respect for the media than me. They are "the" institution we can turn to, to stand up "for the little guy." If there's ever any sort of injustice or a case of someone in a position of power and authority abusing their power, then all it takes is a little media coverage and justice is served, where in my opinion, it wouldn't be served without this media attention. However, what concerns me is the non-balanced reporting of the media. If 99.9 per cent of the military conduct themselves in a professional, responsible, competent manner, then why isn't 99.9 per cent of their coverage devoted to this? Doesn't sell enough papers, I suppose. But the damage is done when the vast majority of the Canadian public have a very limited knowledge of what the military is all about (through no fault of their own) and they will often believe what is in the media. If all the media does is focus on the conduct of a few reprobates, and there are reprobates in all walks of life, then the public is going to get a distorted picture of who and what we are. Enough of my sermon, it's Canadians like you who can make the difference by doing exactly what you're doing, so thanks again!

Well Jane, if I say anymore here I'll probably put you to sleep. For a guy who's a two-fingered whiz on the keyboard and capable of typing only 20 words a minute. this was quite an undertaking. It's great again to hear from Canadians like you, and believe you me, we'll do our best to make Canadians proud. I'm looking forward to further letters from my pen friend where we'll get to know more about each other. And tell your husband and son to get in on the picture! Sure they've got some questions! We might have a lot in common — you never know. I've got some strong Maritime blood in me. Anyway, take care and we'll no doubt be talking again soon.

◆　　◆　　◆

The following letter was written by a 31-year-old captain serving with and writing on behalf of the staff of Alpha Company, 2nd Battalion, PPCLI assigned to the Canadian Contingent of SFOR:

Camp Holopina,
Coralici, Bosnia-Herzegovina
January 26, 1997

Dear Jane,

As the spokesman for this letter, let me introduce myself. I'm a captain and I am the second in charge (OIC) of A Company, the 2nd Battalion, PPCLI. I am 31 years old, am married and this is my second UN/NATO tour.

I say that I am the spokesman as this is a collective response by the staff of "Alpha" Company HQ located in sunny Camp Coralici (near Cazin) in Northwest Bosnia. We've been on the ground here for a month and are still in the process of getting used to our surroundings and learning the nuances of a very big area of responsibility. Before going any further, we'll give you an idea of who we are and how we figure in the great picture of this 1,200 man and woman battle group.

Alpha Company is one of three rifle "companies" that from the sharp end of the 2nd PPCLI. Other sub-units include an armoured squadron, an engineer squadron, an administration company and a combat support company (employing mortars, anti-tank weapons, etc.). As well, a large NSE is located in Velika Kladusa to handle all the logistics required when operating so far from home. A Company is 120 strong and consists of three platoons and the company headquarters and is responsible for the area around the town of Bihac stretching north to the Croatian border and to the south near Kulen Vakuf. It is a huge area to cover and as a result we have a platoon each in Velika Kladusa, Coralici and Bihac. The Battle Group HQ is also in Coralici, so we are in a fairly large camp.

Our task is primarily to monitor the AOR and to ensure the former warring factions abide by the agreements as set down in the DPA of December 1995. Now that's kind of a mouthful, though I am sure you've managed to glean most of the details out of the news. As a NATO force, we pack a bigger punch and a much stronger mandate than that of our UNPROFOR predecessors. This combined with a general sickness of war and killing has resulted in a very quiet time for IFOR and SFOR. We've found that people are generally pretty receptive to our presence as SFOR and as Canadians. Our patrols going into various towns have been invited in for coffee and although struggling with a language barrier, have come away feeling quite welcomed.

Our camp here is very comfortable and there isn't any cause for complaint about the facilities or the food. Many would consider our conditions "austere," however, after having served in various combat arms field units, none of us has a problem getting comfortable here. The only thing that we can't have here is that which we miss the most, our families. As well, once we leave camp, as we often do, it is really a whole different world. Bosnia is very much a Third World country with varying degrees of poverty and large areas of baffling destruction. If every Canadian could see the things that we are seeing here, I suspect none would complain again about Canada.

Your pride and faith in us, is very much appreciated and we hope to continue corresponding with you throughout our tour.

◆   ◆   ◆

The following letter is the third of four written by the naval officer who was mentioned earlier on pages 51-52 and pages 53-55, and who commanded a frigate assigned to Operation Sharp Guard:

**HMCS *Fredericton*,
on patrol, somewhere in the Atlantic
January 28, 1997**

Dear Jane,

I wasn't feeling too hot at Christmas, as I had a bad cold and then ended up with a pinched nerve in my neck and arm for about three weeks. All gone now though!

This deployment has been busy. We are Canada's current contribution to NATO's immediate reaction force of vessels called STANAVFORLANT. We were also part of it in the Adriatic, but this time we are training rather than performing an operation such as the embargo. Although, perhaps not as potentially dangerous, it is much more hectic. Twenty-four-hours-a-day, the ships are exercising at some form of warfare or supporting warfare functions, such as communication exercises, fuelings, etc. Often, I'm only getting a few hours of sleep at a time. In ports we visit, we are a visible symbol of NATO solidarity and have numerous official social functions plus competitions and tours.

We left Plymouth, England, a week ago. While there, I had an average of two official functions to attend per day plus numerous meetings. I only managed to get out and about a couple of evenings and went around with the captains of the

German and US ships. The German [captain] went to school in Plymouth, so knew the area very well. The [American captain] spent Christmas there, so he was pretty much at home as well. Saw some lovely scenery, but it was at least either a week too short for sightseeing or a week too long for meetings!

The end of this week finds us in Hamburg, Germany for three days; then, out for a few days; then, Aarhüs, Denmark; Oslo, Norway; Antwerp, Belgium; Leith and Faslane, Scotland; Amsterdam, Holland; Plymouth, England; and finally Dublin, Ireland; before returning home in mid-April. Just not enough paychecks to cover them all!

Well, I must go. Thanks again for writing. Take care.

◆　　◆　　◆

The following letter is the first of four written by a 35-year-old major who commanded the Reconnaissance Squadron of the 2nd Battalion, PPCLI assigned to the Canadian Contingent of SFOR:

**Camp Maple Leaf,
Zgon, Bosnia-Herzegovina
January 29, 1997**

Dear Jane,

First off, I am married with two children. My wife and I are both 35 years old. Our daughter is nine and my son is seven. I'm living in Edmonton, but will likely be posted to either [CFB] Kingston or Ottawa this summer.

Our mission is now called "SFOR." The name stands for "Stabilization Force" as opposed to Implementation Force." I am what they call a "squadron commander" of a "Cougar" Armoured Fighting Vehicle-equipped squadron. I command a total of 82 people which, I can tell you, is a blast!

I am originally from Vancouver, as is my wife. I enjoy everything to do with the outdoors, but in particular, canoeing and cross-country skiing. I've been in the army for 17 years and am thoroughly enjoying myself with all the various challenges. This country is in a terrible state of affairs. We have a firm grip on the former warring factions' armies, so now we're helping the Bosnians to rebuild their country and survive the winter, but they're finding it bitterly cold. That puts us at an advantage as my regiment (the LdSH(RC)) is stationed in Edmonton and we don't even consider it winter here!

My squadron patrols the entire area the Battalion Group is responsible for which is 8,000 square kilometres and centres on [the city of] Bihac. There's always something different going on here. Militarily, it's pretty quiet as it's winter. Apart from occasionally discovering minefields, we spend a lot of time clearing road accidents and pulling people out of the ditch.

This is my first tour on peacekeeping duties and I'm certainly glad I came. It's very nice to get away from the negativity in Canada and see just what a good thing we really have!

Well, feel free to write anytime.

◆　◆　◆

The following two letters were written by a 43-year-old warrant officer who served with Alpha Company, 2nd Battalion, PPCLI assigned to the Canadian Contingent of SFOR:

**Camp Holopina,
Coralici, Bosnia-Herzegovina
January 30, 1997**

Hello Jane,

I am a 43-year-old warrant officer and have been married for 20 years. I have a son and a daughter and we live in Winnipeg, Manitoba. I have served with the armed forces for the past 24 years and was posted to Winnipeg; Gagetown; NB, Cornwallis, NS; and Victoria, BC; and, back to Winnipeg where I met my wife in 1973. I also plan to retire in Winnipeg.

I am currently serving on my third peacekeeping mission. I was in Cyprus in 1976, Croatia in 1993 and here in Bosnia now; although, this is not a peacekeeping mission, we are under SFOR. We are enforcing the DPA and not wearing UN blue. Our powers of enforcement are much greater than under the blue helmets.

It is very sad to see the spoils of war and the destruction done to such a beautiful country. Especially, to the children and the old. God bless the old. You can see in their eyes, they are tired of fighting. But the young and middle [age] males would start fighting tomorrow, they seem to thrive on pain and misery. After three of these tours, I still can't understand this madness. It is very quiet in this part of Bosnia right now, but with the elections and warmer weather just around the corner, we expect to be busy soon.

Thank you, Jane, for your kind words in support of your CF soldiers, like me, need to hear we still have the support of the public in these trying times and tarnished reputation the Somalia incident has brought upon us who were serving unconditionally, seven days a week.

My morale has suffered with all of the negative media attention given to this wonderful institution, but knowing there are people like you out there taking time to write soldiers is a big boost. Well, you and you family take care and Godspeed.

——————————— ◆ ———————————

Camp Holopina,
Coralici, Bosnia-Herzegovina
February 20, 1997

Dear Jane,

Hi, it's midnight here and I am halfway through my shift in my company command post on a very quiet morning in Bosnia. The atmosphere here is very quiet and people are trying to get their lives back in order.

Over the next few months there is going to be thousands of displaced persons trying to move back into this area for resettlement, and it goes without saying that tensions are going to rise.

Elections are scheduled for mid-July and SFOR is getting geared up for that, but we are not optimistic that they will happen until September. Boy, what a mess!

You are right about the hatred here and a lot of these people especially hate SFOR. I guess its because we won't let them slaughter each other. You will have to excuse my pen scratch and spelling, Jane, but I am one of those people too lazy to use a dictionary, and I am always too much in a hurry to worry about my penmanship. I have to agree with you when you say the fighting is going to start up again. SFOR is scheduled to pull out in 18 months and my guess is they will be back at it again. The boundaries drawn up by the DPA will leave some factions cut off from their fellow ethnic partners. The Serbs are no danger right now, but if I was a betting man, my money would go towards the Muslims being the overall loser when all is said and done.

I wish you good luck on your book and I for one will buy a copy. You spoke of [CFB] Cornwallis and a flood of bad and good memories came rushing back. The "bad" was when I went through boot camp there in 1973 and the good was when I went back as an instructor from 1981 to 1983. It was probably three of the most enjoyable years of my career. It was all topped off with the birth of my daughter.

You may not believe this but we had a surf and turf dinner at the kitchen last Sunday and I had two lobsters in Bosnia. Go figure!

The weather here has been great the past few weeks. It's like May in Halifax, bright and sunny. The snow is all gone, but [there are still] cool evenings. Pretty soon, I'll have to start working on my tan. Ha! Ha! I went to Budapest last week for 96 hours of R&R. It was great. [I] bought some great gifts for my family, had some great meals and a few Hungarian beers. Yum! Well Jane, it's time I get back to work.

◆ ◆ ◆

The following two letters are the second and third of four written by the 35-year-old major who was mentioned earlier on page 79-80, and who commanded the Reconnaissance Squadron of the 2nd Battalion, PPCLI assigned to the Canadian Contingent of SFOR:

Camp Maple Leaf,
Zgon, Bosnia-Herzegovina
February 26, 199

Dear Jane,

As usual, it was very nice to receive your letter. Everything here is going well and the area is quiet, although, there have been burnings of abandoned buildings to discourage the Serbs from resettling. We are dealing with this problem, however.

The locals consider the winter tough this year, but we do not. It's like a mild Halifax winter. They're coping well, however, as there are many aid agencies here who give out food and warm clothing and firewood.

For our part, my squadron has been searching out elderly people living in the hills and checking up on them. We found a few and look after them (some of them are in pretty rough shape). We also keep an eye on the area around them as there are a few young toughs who are not ready to forgive and pick [on the elderly as they] easy targets.

The kids are all back in school (what's left of them). We are working hard in one town to attract aid agencies into providing money to rebuild the schools. Schools were a favorite building for army barracks during the war and were heavily vandalized and looted.

As to whether or not peace will hold after we leave, that remains to be seen. There is an awful amount of hatred here. The civil war gutted the entire country and all three sides did unspeakable things to each other. There are, on the other hand, many people who are tired of the hatred and don't want a return to war. They are becoming very comfortable with the concept of peace.

Well, that's all for now. I hope things are going well for you.

⬣   ⬣   ⬣

Camp Maple Leaf,
Zgon, Bosnia-Herzegovina
March 5, 1997

Dear Jane,

It is my pleasure to take the time to write as it's a form of mental relaxation. You are not wrong, however, things are busy.

Things remain quiet here, but there are political rumblings. Mostar has not affected us, but there are definite problems concerning the resettlement of refugees. Essentially, there is very little work here and no homes for people to live in. I do feel that there is now a chance for "real peace" as the young do not want to fight.

I went for a beautiful patrol today along the Croatian border. The scenery was absolutely fantastic; it's extremely rugged with deep river gorges. We also went up to an old castle built to stop the Ottoman Empire.

By the time you receive this letter, I will be in England for three weeks with my wife and kids. We decided to take my separation leave in England and visit old friends. We'll also be visiting Cornwall (beautiful, rugged countryside), Bristol, Devon, the Moors, Salisbury and London. It should be a superb educational experience for the kids, not to mention a heck-of-a-lot-of-fun. Needless to say, I am counting down the days. Well, that's about all from "Bosnia." Take care.

◆   ◆   ◆

The following letter is the last of four written by the naval officer who was mentioned earlier on pages 51-52, 53-55 and page 78 ,and who commanded a frigate assigned to Operation Sharp Guard:

HMCS *Fredericton*,
at sea, somewhere west of Scotland
March 6, 1997

Dear Jane,

As you can imagine, things continue to be busy. I have just finished a busy and hectic 10-day exercise to the northwest of Scotland. The weather has generally been Scottish (wet, windy and nasty). Winds are constant at 30 to 40 knots, [occasionally] we've had winds [up] to a 60 knots and gusts [up] to 80 knots! That's 140 to 150 kilometres per hour. Really unpleasant. You can imagine how happy I am about that, particularly after my last September's run in with King Neptune!

At any rate, the ship continues to do well and has an excellent reputation with our NATO allies. Since I last wrote, we've done visits to Oslo, [Norway]; Antwerp, [Belgium]; and Leith, Scotland. We're now on our way to Faslane, Scotland for the weekend, then will be in Amsterdam for 10 days. During this time, I will encourage my ship's company (by providing two buses, plus lunch) to make a pilgrimage to Vimy Ridge, Beaumont Hamel and other First World War battlefield sites [in France] where our forefathers created the outstanding legacy which we follow today. I've been there before and it really is moving. Vimy, in particular, is someplace where every Canadian should go to try and understand sacrifice. It was there, that we became a nation among the world of nations. If you haven't read it yet, I urge you to read Pierre Burton's book *Vimy*. It really is an excellent piece of work about a defining moment in modern Canadian history.

Other than Amsterdam, we've got Plymouth, [England] and Dublin, Ireland left. Then it's homeward bound. We're due in on the morning of April 16th. Then I've only got about six weeks left with this marvelous ship and crew. Am I sad to leave?

Oh, to a certain extent, but I've taken the ship about 80,000 miles, have been at sea or away for over 400 days, have done many interesting and exciting (to say nothing of nerve wracking!) things, so it will be time to be home with the family. That I look forward to.

Well, I'll put this in the post! Once again, many thanks for your support

◆　◆　◆

The following letter was written by a 31-year-old master-corporal who served as the Headquarters Company Signals NCO of the 2nd Battalion, PPCLI assigned to the Canadian Contingent of SFOR:

**Camp Holopina,
Coralici, Bosnia-Herzegovina
March 21, 1997**

Hi Jane,

Greetings from another member of A Company Headquarters. Before I get into explaining a bit of my job to you, I will introduce myself. I am a master-corporal and I am a 31-year-old father of two girls who has been married to a wonderful woman for seven-and-a half-years. I "celebrated" my 31st birthday here in-theatre. I use the terms celebrate loosely, as I would have preferred to be home for it. Anyways, you asked the writer of the previous letter how many tours he has had, and I think this is his second. As for myself, this is my fourth trip across the pond. I did 13 months in Germany, six months in Cyprus, six months in Croatia and now six months here is Bosnia. The last three tours were all while I have been married. Actually, I am still married! During the tour in Croatia, my youngest daughter was born, so I missed out on seeing her until she was about three weeks old. It is great to know there are people like you in Canada that support us. This is the first time that I have heard from someone else, other than my wife and other family members of how proud they are of me and the job I do, and you know what? I am also very proud of the work I do, and am very proud of the uniform I wear and the regiment I serve in and am fiercely loyal of Canada. I love my country and just wish there were more citizens of Canada who feel the way you do.

After working with armies from other countries — England, the US, Germany, Austria, France and Italy, I can honestly say that Canadian soldiers are the best trained, hardest working troops in the world. Yes, we don't have the "high-speed gear" of the other countries, but what we have, we use it to the best and even beyond its ability. I just want to say before we go any further is that I am sorry about the spelling and grammar. It is 0400 and I am starting to get a bit tired, not that I am complaining. It goes with the job.

Now, to talk about my job. I am the company signals NCO, which is the senior ranking master-corporal in a rifle company. I am first and foremost an infantry soldier who has been trained in communications, be it through a basic course and then an advanced course (both of which I have done) and then lots of on the job training. My primary responsibility is to provide non-stop VHF communications for the company. This is achieved through using different types of equipment. I am qualified using everything from hand-held radios up to global satellite equipment, as well as fibre optics and a little bit of knowledge in computers (I am still learning) and right now everything is still working, even kit that was built before I was born!

As well as providing communications, I also go out and do equipment checks at our other platoon houses, which means I get out a lot and see the countryside. Unlike the last tour I was on, stuck in the middle of an ongoing war, this time it is very quiet and very safe. The only thing to worry about are the drivers. They drive very fast and the roads are narrow and in poor shape.

Just got off the phone with my wife and she says the snow in Winnipeg, our home base, is about six to seven feet deep and when it melts the worst flooding in a century is predicted for Southern Manitoba. This is weird because the last time I was here in Yugoslavia, in 1993, Winnipeg flooded.

Well Jane, my shift is almost up. Again thank you for writing and it is great to hear from Canadians who feel the way you do about the troops.

◆　◆　◆

The following letter is the last of four written by the 35-year-old major who was mentioned earlier on page 79-80 and pages 81-83, and who commanded the Reconnaissance Squadron of the 2nd Battalion, PPCLI assigned to the Canadian Contingent of SFOR:

Camp Holopina,
Coralici, Bosnia-Herzegovina
May 7, 1997

Dear Jane,

The weather here has been superb. April was terrible with a lot of snowstorms, but now it's very pleasant and the leaves are finally out.

The flood in Winnipeg almost delayed our return to Canada as the Battle Group, which is replacing us, had to stop training to fight the flood. They are back on exercise now, so we'll come home on time.

The people in Bosnia are fine. The World Food Programme ensures everyone has enough to eat and most people get medical attention. There are elderly, isolated people in the hills, but we spend a lot of time making sure they get on okay. There is a lot of house building going on, but it's in select areas. Every government in Bosnia is resistant to people returning home, but the people want to go home. We have problems in our area as the Croats are burning Serb homes and threatening them. Unfortunately, there is very little we can do about that as it is a police/

political issue and we're neither.   The Bosnian Muslim army is acting up in Bihac, so we have a bit of a stand-off with them. The important thing is that we cannot back down and must force them to comply with SFOR. It will be interesting to see who backs down first.

I will be posted to Ottawa this summer. The good side of it is that I have a brother in Ottawa, so I'll be able to get re-acquainted with him. I am due for my last R&R at the end of this month. I'm going to Graz, Austria. I've always wanted to go to Austria, so this is my chance.

Yes, the children are back in school, here. [The soldiers under my command] are actually building a playground for a kindergarten. There are very few jobs here, and that's a problem. Unemployment in the area is at 80 per cent.

Well, until next time, take care! Cheers!

◆ ◆ ◆

The following letter was written by a major serving with the Canadian Contingent assigned to SFOR:

Velika Kladusa, Bosnia-Herzegovina
July 26, 1997

Dear Jane!

What an absolute thrill it was to receive your letter today. I was so very happy to see the envelope as it was delivered to my desk. I knew it was from you immediately. Now everyone knows the story of the great Jane Snailham.

You are right in saying that I've noticed a big difference here.   The shooting and shelling is over and that makes me feel very comfortable. The people are starting to rebuild their homes and their lives and that makes me feel good, too.

You are a special woman, Jane. I'm proud to know you. Five years now! Five years you have written to us in the former Yugoslavia. You are appreciated for your undying support to us. God bless you!

◆ ◆ ◆

The following letter is the first of two written by a major who commanded a Reconnaissance Squadron assigned to the Canadian Contingent of SFOR:

Camp Maple Leaf,
Zgon, Bosnia-Herzegovina
August 17, 1997

Dear Jane,

I am a bluenoser myself. My father served aboard the HMCS *Bonaventure* in

Halifax in the early 1960's. I have many relatives from my mother's side of the family still in Nova Scotia.

My wife and I have two children (a girl, age seven and a boy, age five). My daughter was born in a British military hospital in Germany. My son was born in Kingston, Ontario and my wife is originally from Belfast, [Northern Ireland]. Her father is a veterinarian and they emigrated to Canada when she was six.

I have served in Canada, Germany, England, Africa, the Middle East and now the Balkans. It seems, at times, that I spend more time away from home than my non-military friends, but I am blessed with a loving wife, strong family and quite simply would go crazy in a more "traditional" line of work.

I command a reconnaissance squadron and an extremely proud of the almost 100 soldiers that I work with. During our pre-deployment training, we were ordered to Winnipeg to assist with the flood relief effort and we all found that to be an extremely rewarding experience. With the flood in ebb, we returned to our pre-deployment training. We all managed to get a couple of weeks leave before deployment.

Our job here is interesting. Primarily, we are responsible for preventing the former warring factions from breaking the peace. They are very competent and we have had no significant problems. In addition to our primary role, we assist the international community by establishing a secure environment where aid workers can assist in the rebuilding of communities and resettlement of refugees. This too is going reasonably well: young people are getting married, having children, markets are thriving [and] construction is booming. Even still, Canadians are hardworking people and in addition to our normal duties, our soldiers are visiting isolated elderly people, doing small jobs like fixing doors and windows or simply stopping by for a visit. [Currently,] we are repairing a playground and soon will start distributing school supplies to children. The Canadian soldier is a remarkable person and it is indeed a privilege to command them. In one minute, they can confiscate automatic weapons or detonate mines which pose a threat to others and in the next minute they will smile and wave at children.

In closing, thanks again for your letter of support.

◆   ◆   ◆

The following letter was written by a lieutenant-colonel who served with the LdSH(RC) assigned to the Canadian Contingent of SFOR:

**Camp Holopina,
Coralici, Bosnia-Herzegovina
September 7, 1997**

Dear Jane,

I come from a large family of seven kids. I have one brother in the army, while the rest are spread across the country. My parents live in Belleville, Ontario and I live in Edmonton. I am married with two children.

Things here in Bosnia are going fairly smoothly. We are getting ready for elections. I spend a great deal of time on the road visiting my companies. They are spread over an area the size of Prince Edward Island. I believe it is important to get out and see my soldiers and get their perspective on things. All my travel and other duties keep me busy, so time really flies. We have already been here eight weeks.

Our duties here see soldiers doing vehicle and foot patrols throughout our entire AOR. They do this at all hours of the day and night. As well, they inspect the barracks and armouries of the former warring factions to make sure they abide by the terms of the DPA.

I return to Edmonton in one month for leave. I can hardly wait to see my kids again. One is 12 years old and in Grade 8; the other is 10 years old and in Grade 5. Both are doing well, despite my absence.

I was talking to one of my young captains about your letter. He told me he didn't think I had the time to write as a pen-pal. He also offered to write to you, if you wish. As you can see Jane, I am not timely in the letter writing department — my wife and mother tell me this as well. Anyway, I really enjoyed your letter and will write back to you.

God bless you and your family. Thank you for caring about us all in Bosnia. Your thoughts and words mean so much. Take care for now. Perseverance.

PS — Perseverance is our Regimental motto.

⬡ ⬡ ⬡

The following letter is the second of two written by the artillery captain who was mentioned earlier on pages 74, and who served with the LdSH(RC) assigned to the Canadian Contingent of IFOR, now assigned to SFOR:

Camp Holopina,
Coralici, Bosnia-Herzegovina
October 1, 1997

Dear Jane,

I have been back to work since September 19th. My leave was very good, thank you. I did not do anything exciting on my leave, just a few small trips to Moose Jaw and Winnipeg to visit relatives of my wife. I did a lot of work around the house, like installing window blinds and lying carpet in the living room. We did a thorough fall cleaning of the basement and then had a garage sale to dispose of the junk. Well you know what they say, "One man's junk is another man's treasure." Well it was $170.00 worth of "treasure." Not bad, I thought. My son went back to school (Grade 1) after my first week home. He is in French immersion and seems to like it.

My wife was hired for two jobs on my return to Bosnia. She now works for the Family Resource Centre at our base as a child care worker. As well, she has accepted a job as an on call teacher's aide at the school where my son attends classes. She is very happy to be back to work after six years of "homemaking."

In this letter, you will find a couple of presents. The certificate is something that I thought you might like. I had most of the Liaison Officer section and drivers sign as to their appreciation for your kind words. My boss was the initial recipient of your first letter. As well, I enclosed a Royal Canadian Artillery sticker for you to display if you wish. My grandfather was in the Artillery Militia on the East Coast. He served during the war from 1941 to 1945. Ironic, that I joined the artillery years later, and I did not know my grandfather was a gunner, until 1991. He died in 1982 and I joined in 1990.

As for your questions about the people, here is a little bit of info. Most people have homes to live in. They are not necessarily their own homes because many are refugees and displaced persons from the war. A Muslim might live in a Serbian-owned house, where the Serb had to flee during the war. The homes vary from untouched to completely destroyed. There is much rebuilding occurring by several organizations from different countries. The people work, if they can, and sincerely want to move on with their lives. The children are back to school, but schools are damaged as well. Our rebuilding priorities are schools and hospitals, funded by Canada and Britain. In addition, we have built school playgrounds.

The weather is starting to cool down some. Average temperatures are from a high of 23 degrees to a low of 10 degrees Celsius. They say the snow will stay on the ground by December. It drops to about minus 10 degrees Celsius during the winter. Lately, it has been sunny and warm. No golf courses handy to take advantageof, though. My wife is not quite a "golf widow." She tries to play sometimes. She has her own clubs, but gets too tired to play 18 holes in one day. She would rather bowl anyway.

I am going to sign off now.

◆　◆　◆

The following letter was written by a 31-year-old artillery captain who served with the Canadian Contingent assigned to SFOR:

Camp Holopina,
Coralici, Bosnia-Herzegovina
October 13, 1997

Dear Jane,

I had just arrived back from a long day on a hill near a town called Bos[anski] Krupa. It was about 1930 and I was somewhat chilled after standing in the hatch of my carrier for 70 minutes on the way home. I saw your letter in my basket and knew exactly that it has to be from the "very nice lady" my commanding officer (a lieutenant-colonel) said may write to me. I can tell you it was an absolute pleasure to read your letter. It made me feel proud to know that a great "ordinary" Canadian would take the time to write a good three pager to a stranger. Thanks so much. Well, so where do I start?

I am a 31-year-old captain of artillery based out of Shilo, Manitoba. Back in Shilo, I am the battery second-in-command for "A" Battery. (An artillery regiment has four batteries.) Over here in Bosnia, however, I am a Forward Observation Officer/Forward Air Controller (FAC). My job is to control artillery and anything that comes "down" from the sky, (i.e., jets with bombs). I spend most of my time "FACing" which is controlling NATO jets on simulated attack runs. I have a crew of three and work out of a "Grizzly" armoured personnel carrier. I love it! Today, at Bos[anski] Krupa, from 0800 to 1800 hours, I talked to four U.S. F-16 [fighter/ bombers] and four French Mirage fighters. What a blast! My guys are enjoying it too. We have also worked with British artillery that is stationed over here and have adjusted the fire from their guns. This may sound somewhat war-like, but our job over here is to keep the "former" warring factions just that — former — and we have to be able to backup our talk if required. I honestly hope, it never comes to that! I enjoy my work and that camaraderie, but I miss my family. I have been married for seven years. [My] wife is a constable in the Royal Canadian Mounted Police. We met in Gagetown while I was going though training. (This was her first posting). She is absolutely gorgeous and the best person I have ever met. We have two beautiful boys, aged four and six. I was just home on leave in September and it was tough saying good-bye. I won't see them again until sometime in January. We talk a lot (free phones over here — great eh!) and I write. She is very busy with the boys, work, two research projects at university and chairpersonof two boards on the base! Ambitious, eh?

Getting back to the army, life has been pretty good over here, so far. The people (Muslims in my area) are quite friendly and seem glad to have us here. There is a lot of rebuilding going on, but the winter should be settling in soon, so that will slow. The fall has been absolutely fantastic, thus far. Just like a Maritime fall except notas colourful yet. Well I've rattled on. Never let a Maritimer resides in Nova Scotia talk about work and family in the same letter, eh! Ha Ha!

Again, I think it is fantastic that you are writing to myself and other peacekeepers. You sound like a fine person and a great Canadian and I hope we can keep up the dialogue. All the best to you and your family and do take care.

PS — Thanks also for the card!

◆   ◆   ◆

The following letter is the second of two written by the major who was mentioned earlier on page 86, and who commanded a Reconnaissance Squadron assigned to the Canadian Contingent of SFOR:

Camp Holopina,
Coralici, Bosnia-Herzegovina
October 17, 1997

Dear Jane,

I have just returned to Bosnia from my mid-tour leave. It is 0300 here, 0700 at home in Edmonton, and I am using my jet lag to catch up on paperwork that has accumulated in my absence. It was a pleasant surprise to find two letters from you amongst all the situation reports in my basket.

For leave, I spent the first five days in Edmonton, then my wife and I took the children to Disneyworld in Florida for a week. My daughter will have many fond memories, but my son at five years old will probably forget the trip as he grows up. The important thing, however, is that we had some wonderful quality time as a family. We celebrated Thanksgiving with a friend who recently had her third child and whose husband, although, home for the birth, had to return to Bosnia the week before Thanksgiving. It was an excellent Thanksgiving dinner.

On the humanitarian side, my squadron is very active here in Bosnia. Keep in mind, that enforcement of the DPA is our primary mission and that we are one of four sub-units in the Battle Group. The others do a similar amount of humanitarian projects.

We regularly visit over 60 isolated elderly people and provide firewood, food and other necessities of life. We keep the humanitarian organizations appraised of their needs and sometimes, simply provide company to people living on their own, miles from their nearest neighbors.

We have just finished constructing a playground at one elementary school and will begin another as soon as the engineers clear the land of mines and booby traps.

We are assisting a humanitarian organization with the construction of a cultural centre in one of the towns near our camp — they provide the materials and we provide the labour.

I have a plan at Christmas to mount a small expedition to a community of elderly Serbs who got cut off in the mountains, due to impassable roads when the snow flies and I am certain that our soldiers will find many worthwhile needy causes to volunteer their spare time to assist.

Must close for now, thanks again for writing.

⬡　⬡　⬡

Cambodia

The Canadian Contribution to Cambodia

Mission Background

The history of Cambodia is a one of tragedy. For several centuries, it was the centre of an empire encompassing most of Southeast Asia and later a French protectorate from 1863 to 1954. Granted independence by France in 1954, Cambodia's foreign policy of its royalist government was one of strict neutrality. However, this fell on deaf ears as the war in neighboring South Vietnam spilt across its borders. Throughout the 1960s and early 1970s, Cambodia served as a conduit for troops and supplies from North Vietnam destined for South Vietnam. Cambodia's leader Prince Norodom Sihanouk permitted North Vietnamese regulars and Viet Cong guerrillas to operate from Cambodia's frontier provinces adjacent to South Vietnam as a sanctuary from American and South Vietnamese forces in hot pursuit. Consequently, this issue divided the central government into two political camps: neutralist and anti-communist. In 1970, while Prince Sihanouk was vacationing overseas, the military led by former Defence Minister General Lon Nol staged a *coup d'état* which installed an anti-communist regime.

Subsequently, American and South Vietnamese troops invaded Cambodia to destroy the communist sanctuaries and later withdrew following an international outcry. At the same time, disgruntled peasants led by Maoist revolutionaries under the tutelage of Pol Pot formed the communist Khmer Rouge (KR) and fought a bitter five-year civil war from 1970 to 1975. Cambodia became one of the most bombed nations in Southeast Asia (from 1970 to 1973) as American B-52 bombers and military assistance attempted to prop up the central government, but to no avail. In April 1975, the KR triumphantly entered the capital of Phnom Penh.

The KR set about building an agrarian-based society by emptying the cities and did not spare any life to reach this end. Individuals who were perceived to have even the slightest of an education were brutally tortured and murdered. From 1975 to 1979, an estimated two million people died in what was to be known as the "killing fields" of Cambodia. The genocide ended when boundary disputes with Vietnam led to armed skirmishes in 1977. In December 1978, Vietnamese troops launched a full-scale invasion of Cambodia. Vietnam's occupation of Cambodia was complete by early 1979.

What followed was 10 years of foreign occupation and renewed civil war. In 1979, the UN called for the withdrawal of all foreign forces from Cambodia. In 1982, a government-in-exile was declared by a loose coalition that included supporters of the KR and the former royalist government. Most countries refused to recognize the legitimacy of the Vietnamese-backed government in Cambodia. In 1983, the UN recognized the coalition-in-exile as the true government of Cambodia. While fighting continued between the Vietnamese and various communist and non-communist resistance forces, in 1987, the Soviet Union and China led efforts to get the various factions to negotiate a cease-fire. A year later, Vietnam signalled its intention to withdraw its forces from Cambodia. The majority of its nearly 200,000 occupation troops were withdrawn by late 1989.

In January 1990, Australia put forward a peace proposal which was accepted by the UN Security Council. This led to a series of negotiations, which in turn led to a fragile cease-fire and the approval (in 1991) of UN Security Council Resolution 717 which established the United Nations Advance Mission in Cambodia (UNAMIC).

United Nations Advance Mission in Cambodia (UNAMIC)

The Cambodian peace accord was signed in Paris in October 1991. Canada was among the signatories and agreed to deploy CF personnel as part of the approximately 400 troops from 23 countries that were assigned to UNAMIC. UNAMIC was established in October 1991. The initial Canadian Contingent consisted of logistics personnel, engineers, administrative support staff and staff officers who were involved in arranging communications relating to the cease-fire, establishing mine awareness training and de-mining programs to enable the resettlement of almost 400,000 refugees.

United Nations Transitional Authority in Cambodia (UNTAC)

Despite sporadic fighting in parts of Cambodia, by 1992, the UN felt UNAMIC had achieved sufficient stability in Cambodia to move to the next phase of the peace plan: establishment of the vastly larger UNTAC.

The mandate of UNTAC, which eventually fielded 20,000 military personnel from 20 countries, was twofold. First, the mandate called for the supervision, verification and monitoring of the cease-fire as well as the disarmament and withdrawal of foreign troops, including making sure no foreign troops or clandestine military assistance re-entered Cambodia. Second, the mandate provided for the supervision of democratic elections held in May 1993. With the creation of a new Cambodian government on September 24, 1993, the UNTAC mandate concluded.

In March 1992, the Canadian government announced Operation *Marquis*, under which 240 CF personnel would deploy to Cambodia. This contingent consisted of engineers, a transportation logistics unit, a contingent of naval observers and headquarters personnel.

In an attempt to deal with the estimated five to 10 million mines remaining throughout Cambodia, UNTAC began efforts to conduct mine awareness programmes, mine clearance training and the planning of mine clearance operations. Although UNTAC's mandate has since been terminated, these efforts continue under the auspices of the UN Development Program Mine Technical Advisory Group to the Cambodia Mine Action Centre (CMAC). Seven CF personnel are currently assigned to the CMAC.

Canada's naval observers served as an important contribution in executing the mandate of UNTAC. As with all the UN forces in Cambodia, the role of the naval observers in general, was to supervise the cease-fire, confiscate weapons from the various factions and ensure no additional military forces or weapons entered the country. UN naval observers were also assigned the task of establishing a registry system for all commercial-sized Cambodian boats.

Among the 30-strong contingent of Canadian naval observers Canada's Maritime Command assigned to UNTAC, seven were located with UNTAC headquarters in the capital of Phnom Penh and nine were assigned to Ream, a naval base on the Gulf of Thailand. The remaining 14 were divided into groups of two or three serving with various river patrol and border contingents. All the Canadian personnel were rotated at monthly intervals (or more often from locations where the climate and heat were severe).

Prior to leaving Canada, all CF personnel who volunteered to serve in UNTAC received at least a week of briefings and specialized training. These sessions incorporated some of the routing requirements of any peacekeeping operation, such as inoculations, medical and financial arrangements but also provided information specifically about Cambodia. Canadian naval observers attended briefings on the geography, climate and history of Cambodia as well as information on the culture, lifestyle, customs and habits of the Khmer people. Canada's bilingual nature served as a bonus to UNTAC, since most Cambodians, due to their colonial legacy, speak French.

Initially, most patrols were conducted only during daylight hours. Although some of the Cambodian vessels used by Canadian naval observers carried light weapons, observers were restricted from carrying small arms unless the situation became untenable and their safety became a matter for concern. UN river patrols plied the entire length of the Mekong River, including several other rivers that flows into it, and Tonle Sap Lake and the river that flows from it, connecting with the Mekong River at Phnom Penh.

Half the Canadians carried out staff and support functions, which included the maintenance of vessels and other equipment and trained Khmer (Cambodian) personnel. The rest performed observer duties. Typically, the captains and crews who manned the American- and Russian-made patrol boats were Cambodian, with Canadian and other UN personnel aboard as mechanics, advisors, trainers and observers.

This section is based on public information documents produced by the Canadian Department of National Defence office of Public Affairs.

Letters from Cambodia

The majority of the following letters detail the correspondence between Jane Snailham and one Canadian naval observer who served in Cambodia from July 1993 to August 1993. Wherever the location of the peacekeeper writing the letter is unknown the location is noted as "somewhere in Cambodia...."

The following two letters were written by a 26-year-old corporal who served with B Platoon, 92nd Transportation Company assigned to the Canadian Contingent of UNTAC:

Somewhere in Cambodia
January 19, 1993

Hello Jane,

Well, to let you know a few things about me. I'm 26 years old, separated for one year from a down-home girl I was married to for five years. I've been in Cambodia since October 1992. I am leaving to go back to Canada on April 25th. There are about 200 Canadian peacekeepers here from all over Canada. Ninety per cent of the soldiers here are from CFB Petawawa, including myself. Our taskings here are to set up for the elections in May. So far everything is going good. The only thing that really gets to us is the heat that reaches 45 degrees Celsius and higher, at night it goes down as low as 22 to 28 degrees Celsius which in Canada that's hot even for midday.

This country is still war-torn; they're trying to fix the areas that were badly destroyed after 20 years of fighting. Our main job here for Canadians is transport; we deliver supplies, food and equipment all over the country. Sometimes we leave here and not return for three to four days at a time.

This country is the same size as Newfoundland, but with the roads as they are, 100 kilometres could take us six to seven hours. We are not allowed to drive at night because that's when the KR does their attacks on the villages and towns.

Besides all that, everything here is not too bad. Just being away from family and friends at Christmas was the hardest, but I have family here also. All the soldiers here are very close, so that helped over the holiday, and letters from people like yourself really helped. I must let you go, duty calls.   So take care and once again, thanks for your support.

——————————— ◆ ———————————

Somewhere in Cambodia
March 27, 1993

Hello Jane,

How are you doing? Hope everything is going alright. I just got back from an eight-day trip up north, which was only supposed to be a two-day trip. I received a couple of letters. It's always nice to hear from you. Sorry I couldn't write sooner. So you guys are having a bad winter, eh? Oh well, it will be over soon. The weather here is still the same. The rainy season is creeping up on us slowly, last week we had three days of rain. That was the most we had in the last five months.

Can't wait! Twenty-seven days left, then we leave and get back to some good weather. We were scheduled to fly to Guam, then to Hawaii, but it's all canceled now. We are flying to Thailand, Japan, then to Alaska, then Canada. It takes three days. We should arrive in Trenton on April 28th, then we have a three-hour bus ride to base. It's going to be a long week as soon as I get to Canada. I'll be going on leave for awhile after we clear into the base. I don't know if I am flying to New Brunswick or driving, but if I do drive I will try to stop in on the way down. If not, I'll drop a line or two, to let you know.

We've been pretty busy here the last little while. The elections are getting closer and a lot of people out of the four factions that are running for power are causing a lot of trouble. The presence of the Cambodian army and the KR soldiers are firing a lot more shots at each other, not too many casualties, just a lot of useless gun fire into the air to scare the other force. It's going to get a lot worse within the next four to five weeks. I just wish it will go okay. Everyone knows there is going to be trouble.

The Vietnamese that are here are starting to go back to their own country because they are scared of the KR. If they are elected, they will get rid of all the Vietnamese people, the same as they did in 1975. Oh well, it's getting late and I'm running out of things to talk about. Take care and I'll drop a line later.

◆ ◆ ◆

The following five letters were written by a lieutenant-commander who served with the naval observer mission assigned to the Canadian Contingent of UNTAC:

Sihanoukville, Cambodia
July 2, 1993

Hello Jane,

I have been married to my wife for 26 years and we live in Dartmouth, [NS]. We have two sons. Our eldest is 25, a graduate of the Royal Roads Military College and he is in the Air Force. He is married to a young lady from Kelowna, BC and they have an eight-month-old daughter who is, needless to say, the apple of her grandparents' eyes. My son and his family have been posted from [CFB] Goose Bay to CFB Borden. After three years in "the Goose," they are looking forward to civilization. Our second son is 23, in the Army and stationed in Calgary. He is still single (I think).

I joined the Navy in September 1964 and have had a great time. We have been on two exchange postings, one to the UK from 1978 to 1981 and the other to Norfolk, Virginia [USA] from 1985 to 1988. I served two months in the Persian Gulf during the war over there and this is my first UN tour.

There are 30 naval types over here. I am the Contingent Commander for them as well as the Coastal Commander in Cambodia. The training package we did prior to our departure was very good. We did a full first aid/CPR [cardiopulmonary resuscitation] course, small boat driving course which qualified us as small boat coxswains [a sailor who steers the boat], range work, self defence boarding and general policing info, plus lectures from National Defence Headquarters. It was all quite good and has proven useful.

We left Halifax on June 4th at 1800. We flew to Dorval, [Quebec], a bus to [Montreal International Airport] in Mirabel, [then flew] KLM [Airways] to Bangkok, [Thailand] overnight there, then Cambodian Air to Phnom Penh, arriving noon on June 7th. What a groaner that was! The trip from Amsterdam to Bangkok was [over] 10 hours long and took us over Bonn, Moscow, India, Tibet, parts of the Middle East etc. It was mostly at night, but what I did see was pretty fascinating.

The arrival in Phnom Penh was in the midst of a real scorcher. It was around 40 degrees Celsius. Then came the ride of a lifetime. The trip into town was unreal. There are literally thousands of motorbikes and smuggled cars. People don't have licenses and there are no rules of the road. The term "might is right" took on a new meaning here. We had to take a UN driving test and proved once again that Canadians can take on anything and do it well.

My job is the Coastal Commander stationed in [port of] Sihanoukville in the south of Cambodia on the Gulf of Thailand. I have five outstations that carry out river coastal patrols in their area. I am responsible to Naval HQ Phnom Penh for everything that takes place navy-wise on the coast. Basically, I am the filter. The base at Ream [near Sihanoukville] is the largest of the outstations and I am the OIC. I have a [British] operations officer and a New Zealand chief working with me in the HQ. Right now, we are preparing for the UNTAC withdrawal. It is very quiet in here. Last week, we had a major anti-smuggling operation where we inserted marines by air and pushed things to the limit. The Force Commander was impressed by the operation and the squadron commander said although it went well, I had been pretty close to the line. I knew that, but my biggest concern was the safety of the personnel. Fortunately, I won!

The weather has been exceptionally good over the last couple of days. The rains seem to come at night and early in the morning. I'll tell you one thing, the thunder clappers are really horrendous. They usually take out the power, and of course, the water build up is tremendous because of the lack of proper sewage/ drainage system. It is extremely hot and humid though. The people are very friendly towards us. They are very resilient and seem to cope well with their extreme poverty. I have seen some sights that would rip your heart out and there is nothing we can do.

So far things are not quite as I thought they would be. As I said earlier, it is very quiet and I am not all that busy, although, I must be in the office every day, just in case! I plan on taking Saturday and Sunday afternoons off. I will be on the hand-held radio if I am required.

Once I have enough time over here, my wife intends to come over around the first part of October. We will meet in Bangkok, stay there for a couple of days, then go to a resort in the south [of Thailand] called Phuket. We will then come back to Bangkok for a few more days. Hopefully, this will all take place without a hitch. It will be a real eye opener for my wife. In all my travels, I hadn't seen sights like that anywhere. But still, the people press on.

Well, we had a Canada party at our home last night. The guys did a real great job. They put together about 150 to 200 kabobs of beef and veggies, plus shrimp. The barbecues were scrounged. There were tiny lights strung up and everyone who attended is still talking about it. Oh yeah, we also had a Canadian flag made in the market. It just added a little bit of home.

That's about all for now, Jane. Again, I appreciated the letter and will be glad to continue in the future. Take care of yourself and stay happy.

Sihanoukville, Cambodia
August 10, 1993

Hi Jane,

Received your letter yesterday, thank you very much. Is the weather as crazy this year as it was last summer? It certainly sounds like it.

Well, I certainly had a very interesting, in fact, great weekend. It was a trip for a military writer who was doing an article on the world's navies in peacekeeping. He was to arrive at Ream Air Base at 0800 on Saturday, August 7th. Quick as a flash, the helicopter was over an hour late. When we finally sailed, the weather started off okay, but quickly deteriorated into rain showers and rolling swells. That sure didn't help the two naval pilots, the UN Naval Observers or the Australian lieutenant-colonel reporter. On top of all this, the CPAF (Cambodian People's Armed Forces) cook was preparing lunch and the smell was rancid! That really put the kibosh on these guys. The cook was preparing something [that looked] like a deer or mouse. Hanging in the gallery, it looked more like a skinned dog!

The coastline down around Kep and Kampot [east of Sihanoukville] is certainly different than around Ream. There are more mountains and it looks like a jungle

area. It reminded me of the movies you see about the Vietnam War. Having left around 1000, we arrived off the coast at about 1330. We transferred to a rubber boat and were taken into Kep. This is where another Canadian took over as team leader because of all the problems I had down there. Just another story in Cambodia! We had a good briefing, then a petty officer took the lieutenant-colonel on a tour of Kep. [Kep] is really isolated and surrounded by the NADK [National Army of Democratic Kampuchea — the military wing of the KR], CPAF and SOC [State of Cambodia] police. All of who don't like each other, let alone UNTAC. We were then picked up by another Canadian and a couple of his lads and transported to Kampot which is only about a half hour drive on one of the worst roads.

Kep and Kampot were the bourgeoisie places during the country's height but were completely destroyed by the KR. It is quite evident what this must have been like. Anyway, arrived at the navy house which is very large, but no air [conditioning], so very hot. We sat around having a few beers and answering the reporter's questions. The meal that was served was humongous! There was everything from crabs to spring rolls, to ribs, to shrimp etc. After supper, the stories continued and we hit the hay around 2200.

There was lots of rain during the night and I heard an AK-47 being fired — they do that to stop the rain! The roads in Kampot are absolutely atrocious. They were flooded to the extent that the land cruiser was pushing water with its headlights. We then went down to HQ for a briefing and then off on a sea/river patrol. We went down the river to see if smugglers [who] were aground from the night before were still there. They were off loading their cars just down from the river base. After reporting this, we took off up the river. It was really pretty in spite of the downpour! We stopped at a number of villages to treat the kids. It was something to see the way people have to live in these places — no conveniences what-so-ever! Sometimes the filth could get to you. The kids were always smiling and waving and racing down to the river's edge to see us. I really did enjoy that.

After we finished that patrol, we went on a vehicle patrol into some really remote places. The kids come scrambling out of their huts when they hear the vehicle and they're yelling UNTAC! In one place, three little ones came running through the water in front of their place and the youngest, about three, whips her dress up so as not to get it wet. Then she almost drowned because the water was so deep! The brother hauled her up. The smiles were all over their faces. It made you think how greedy free world kids are! They would only get one candy, but they were so thankful for that. The PO said that no matter how badly things get, the kids make up for it and I can see why! There were a couple treated for cuts and bruises and before you knew it, it was time for lunch. The lieutenant-colonel and I paid for lunch. For 12 people, it came to $35.50! Later, I was driven back to Sihanoukville. It was a great weekend and I hope to get some of the others down before it closes.

Yesterday did not start off all that well. My chief petty officer received a call from New Zealand from his wife, informing him of the death of his mother. Then his dad called. It was expected, as she had cancer and he had been home oncompassionate leave in June. It still doesn't make it easy.

Headquarters called at 1700 and told me the CPAF pay master was coming down with 91 million Reil [the Cambodian national currency] to pay the sailors and I was to have everything arranged! Did it ever tick them off when my reply was, "It has already been done." So efficient.

An opportunity came for my operations officer to go to Kep and Kampot, so off he went early this morning. There are a thousand stories in our days and these are but a few of them! This is what makes this so interesting!

Must go for now. Hope this letter was of interest. Take care of yourself and enjoy your summer.

Sihanoukville, Cambodia
August 19, 1993

Dear Jane,

Things are going very well over here as we wind up to wind down. Thank you very much for your card and picture.

To answer a few of your questions, we do have a lot of contact with the local people. As I mentioned in one of my letters (I hope), six Canadians live in Canada House with a family, although, they live in the back part of the house. When I get back, I will bring over my pictures and tell you some of the stories, but let me give you a run down on the last few days.

Saturday the 14th, was a very rainy day. My coxswain called to say the PO's son was on life support, lungs filled with blood and basically brain dead. Expected the family would make the decision to pull the plug very soon. Pretty hard decision! My PO was one of the team and was flown home on August 9th when he found out his son had been in a car accident.

Sunday started off sunny, quiet and I was alone as I gave the PO and operations officer the day off. Went back to Canada House early. Then, the Canadian team leader, called on the radio from Kep. Basically, a CPAF soldier had been shot through the head by a "Vietnamese" fishing vessel that we think the CPAF were probably trying to board to extort fish. The CPAF had brought the body to the team leader to explain the story and have his team take the body to the family. The only way I would allow this was to have the CPAF absolve the team of all responsibility and that they handle the body. This was agreed to, and all was finished with that. Then, as I lay reading around 2100, the radio started again. This time it was another Canadian team leader who was dealing with smugglers. Customs and SOC police went off with the team to investigate an incident. They found a boat loaded with cars and directed them to not unload. Then Customs and the SOC police wanted the UNTAC team to post a vehicle overnight! The team leader asked and I said not unless you have protection. The French Battalion had already departed on their patrol and rightly so. That ended that.

Monday, I was off on an overnight patrol toward Kep and Kampot as all the excitement and killings were taking place down there. My decision was to put more of an UNTAC presence down there. Just as I was about to leave, I got the word that the PO's son had passed away. Then we sorted out our donations, headed out on the 1400 class [patrol boat] under bright sunny skies. Had a very good trip down, but did not see any targets of interest. Convinced the captain to go alongside in Kep at 1630 as we were off duty again at 0100. The patrol leader and I went up to Kep House for dinner with the lads. Met with the CPAF colonel of the area to discuss the killing and the extortion going on. Interesting conversation. When he departed, had noodles in spaghetti sauce which was great. Had a little sing-a-long, then hit the hammock for two hours. Sailed at 0100 and did a saw tooth patrol, north and south, just touching the Vietnamese border before turning north. Did not see anything! Probably, because when we arrived, there were lots of fisherman who saw us and the word was out — UNTAC and myself are in the area! The night was beautiful and when we headed back toward Ream, just as we came over the top of Dao Pho Qouc Island (Vietnamese-held), we ran into a couple of targets. Although we did not board, we did illuminate. Got in around 0800, as the weather had deteriorated. I was impressed with the Captain's seamanship abilities and his cooperation. I let him know through our interpreter. He seemed pleased. Arrived at Canada House, showered, changed and was back at work by 1030. By 1630, I was knackered, so went back to the house at 1700. Hit the rack at 2040 and didn't move until 0550! Tired or what!

Wednesday was quiet and I managed to get some stuff done. It is Personnel Evaluation Report time, plus the administration stuff for end of mission is starting to flow, so my weekends will be bought up with that.

There is still a lot of speculation as to what is going to happen at the end of UNTAC. (Whenever that is?) Might wind up singing Jingle Bells here, especially, the way the civilians are going.

Looking forward to October when I get my time off and see my wife. It will be 115 straight days of work. I sleep with my radio as well, so sometimes there is a lot going on during the night as well. Like I said — each day is different and interesting.

That's about all for now, Jane. Please take care yourself. Write again soon.

Sihanoukville, Cambodia
August 19, 1993

Dear Jane,

Well here it is, a rainy Saturday and I am plugging along attempting to get some final administrative things done prior to going on leave with my wife. We are looking forward to this vacation so very much. It will, in fact, be the honeymoon we never had.

As I was sitting out in front of Canada House this morning at 0545 having a coffee, I started to think about how fast this mission is coming to an end and about some of the things that have happened. So, if you will bear with me, I plan on rambling along for the next little bit. Thanks in advance.

I have yet to see the moon since I have been in Kampong Soam [west of Sihanoukville]. We can't really figure out why as we have had some glorious evenings. With all the mosquitoes around, they don't make that annoying buzzing sound. They just nail you! Bed bugs are just as bad. You never see them but once they attack look out!

The kids here grow up so fast. They don't appear to have a childhood. It's nothing to see a four or five year old with a little one on her/his hip toting her/him around. They play with them and feed them. The kids also do things without being "nagged" at — wash vehicles, sweep the yard, clean up, etc. It is an everyday requirement for them. Very seldom do you hear them crying either. I can count on two hands the number of times I have heard a child cry. They are usually smiling and waving especially at the UN vehicles. They are so gracious when you give them money or candy.

Motorbikes! The ultimate in Cambodian transportation! It is amazing what you see being carried on these little bikes. There have been the front frames of trucks, (no kidding), large window panes, huge bags of grain, other bikes, insulation, chickens, ducks, kid's families, (the record is eight on one bike), lumber and bricks — basically anything we haul in a pick-up truck, they do on these bikes. The most bizarre one I saw, and I couldn't get my camera out fast enough, was two pigs tied sitting up in the front of the handlebars so it appeared they were driving it! There are literally thousands of them around. I have seen women combing their hair, sorting groceries and even nursing! All without clutching for dear life as I would be.

Animals wander around this country just as if they were wild. It is nothing to see a herd of cows or pigs wandering down the road toward you. At night, you have to be so careful as the cows will lay down anywhere to rest! It is so black (no moon) you don't see them until you are on top of them. Fowls, dogs and others are just as plentiful. Speaking of fowls, the roosters have no concept of time in this country. They start crowing any time after 0200. This gets the dogs going, then the KR start shooting and the cycle goes on! Speaking of shootings, some of the KR believe firing into the air will stop the rain or winds or any other environment occurrence that does not appeal to them. If you don't like it, shoot it!

It will be nice to drive around the countryside at home and not see armed soldiers with AK-47's or B-40 grenade launchers. Also, it will be great to see pick-up trucks that don't have 50-calibre machine guns mounted on the back. I look forward to walking legally anywhere in the country without having to worry about mines. It is a real sin too, because the majority of this country is quite beautiful and it would be nice to go up into the hills or into the forests just to explore (not to explode.)

The rains over here are not as depressing as they are at home. You know how we complain after only a couple of days of rain — over here it is a way of life. Ten to 12 days at a time is not unusual and once accepted, it doesn't phase you.

Clean water — the commodity we take for granted at home. I will never waste it again. It is something everyone has to think about. You have to see what people bathe, cook, wash clothes in or swim in, to appreciate what we have at home. The kids over here thoroughly enjoy the rains and to see the amount of fun they have in just puddles with nothing but a stick to play with — resilient or what! Speaking of things to play with, I have seen some pretty interesting toys made from nothing but cardboard, tin cans, hunks of wood, string, paper, leaves, etc. There are no high tech toys over here folks!

Communications is another thing that can be a source of frustration if one lets it get to you. I have been just trying to contact Phnom Penh by phone for the last couple of hours to no avail. Helicopters that are supposed to bring mail — official and personal — have a tendency not to fly or turn around because of "bad weather." Reports you need don't arrive and things you send up to Phnom Penh take forever to get there. Mail from home is the biggest morale booster and when that doesn't get through people get pretty miserable.

Sihanoukville, Cambodia
September 21, 1993

Dear Jane,

WE SAW THE MOON LAST NIGHT — ONLY A QUARTER BUT IT WAS THERE!

One of the great things, and there are many, about this mission is the number of different nationalities I have had the privilege to work with: Australians, Americans, Bulgarians, Brits, Bangladeshis, Canadians, Chinese, Chileans, Dutch, French, Ghanaians, Irish, Indians (East), Indonesians, Japanese, Kenyans, Khmers, New Zealanders, Poles, Pakistanis, Russians, Thais, Uruguayans and Vietnamese — just to name a few. Everyone has their strengths and weaknesses and some nations are less tolerable than others!

The adventure has been tremendous and I am so glad I had the opportunity to participate. It is safe to say that the UN has made a difference in this country. I am not sure that it is as great as the money spent, but a little, especially in a third world country, is an awful lot. The changes I have noticed in my area of the country, and the odd time I get to Phnom Penh is very evident. We can only hope that it continues.

PS — By the time you receive this, my wife and I will be together in Thailand after her whirlwind tour of Hong Kong and Singapore. I should be home the third week of November. I will be in touch and look forward to meeting you.   Take care.

◆　◆　◆

Map 1. Canadian Contingent troop dispositions in UNPROFOR, March 1995. Map not to scale.

Figure 1. July 1992: a UN convoy moving towards Sarajevo. (author's collection)

Figure 2. Canadian peacekeepers burning clothes and linen at Fojnica Hospital. (DND photo)

Figure 3. A Croatian woman jokes with a Canadian soldier at a checkpoint.   (DND photo)

Figure 4. Canadian United Nations peacekeepers patroling a devastated Bosnian town.   (DND photo)

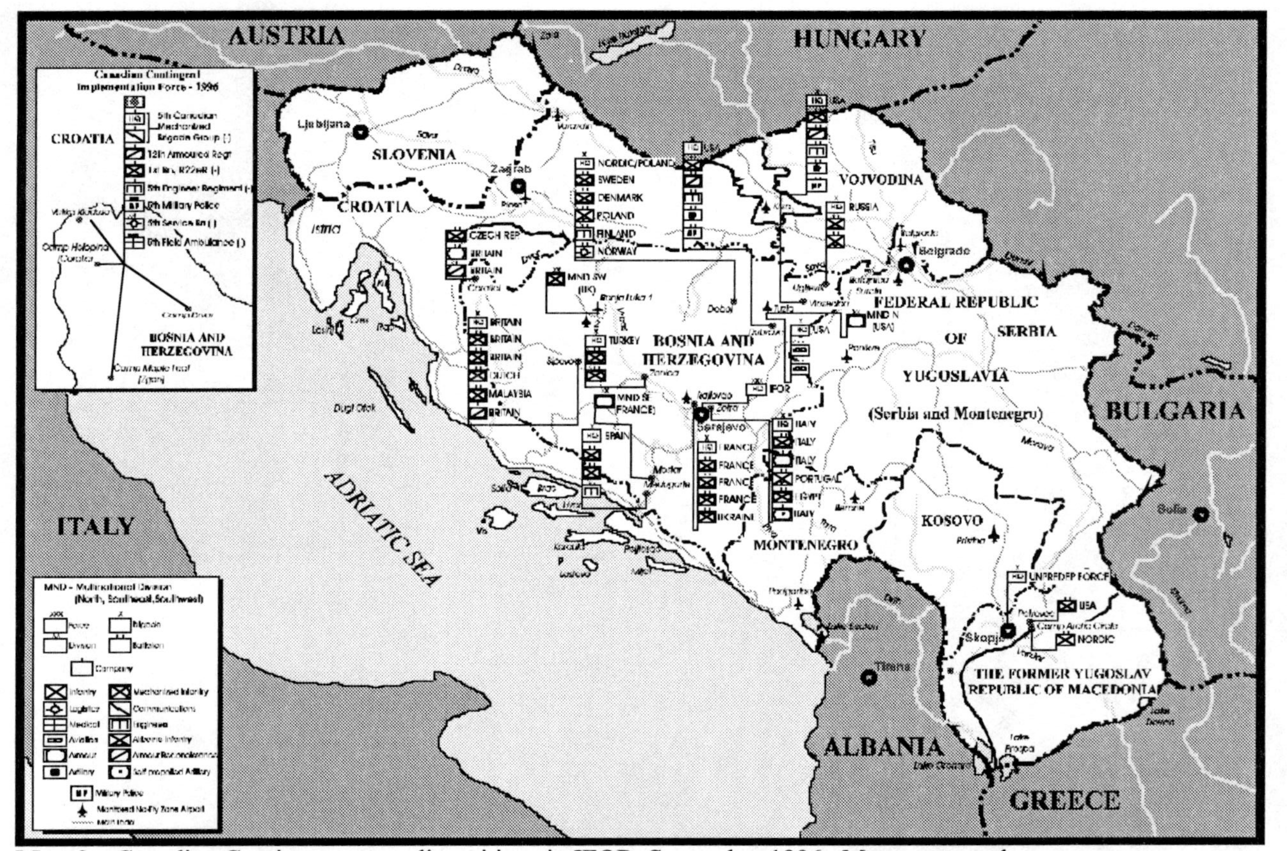

Map 2. Canadian Contingent troop dispositions in IFOR, September 1996. Map not to scale.

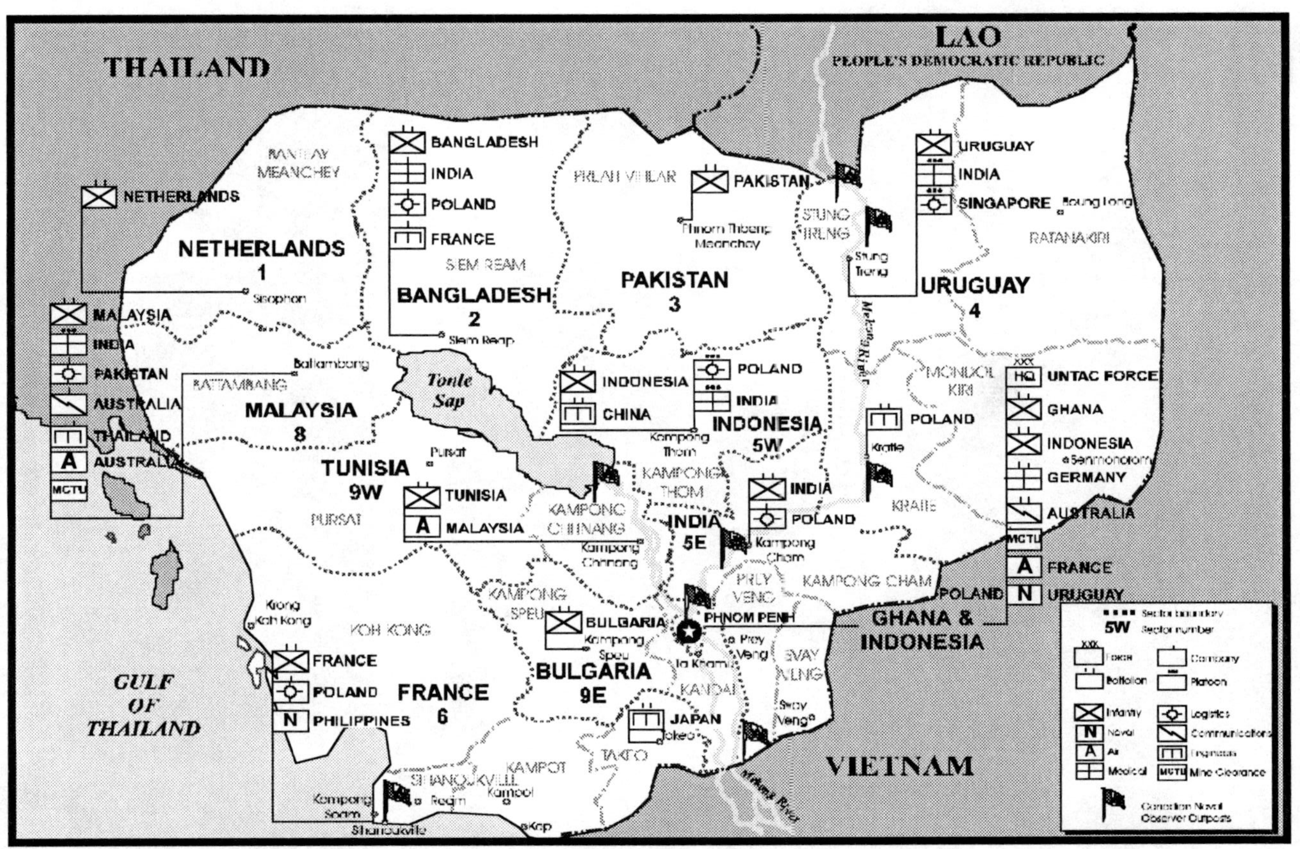

Map 3. Canadian Contingent troop dispositions in UNTAC, May 1993. Map not to scale.

Figure 5. Cambodia 1992: Canadian soldiers distributing toys to local citizens.

(author's collection)

Figure 6. Canadian medics with UNTAC treat malnourished Cambodian children.

(author's collection)

Figure 7. Canadian "hearts and minds" at work in Cambodia distributing aid and assistance.

(author's collection)

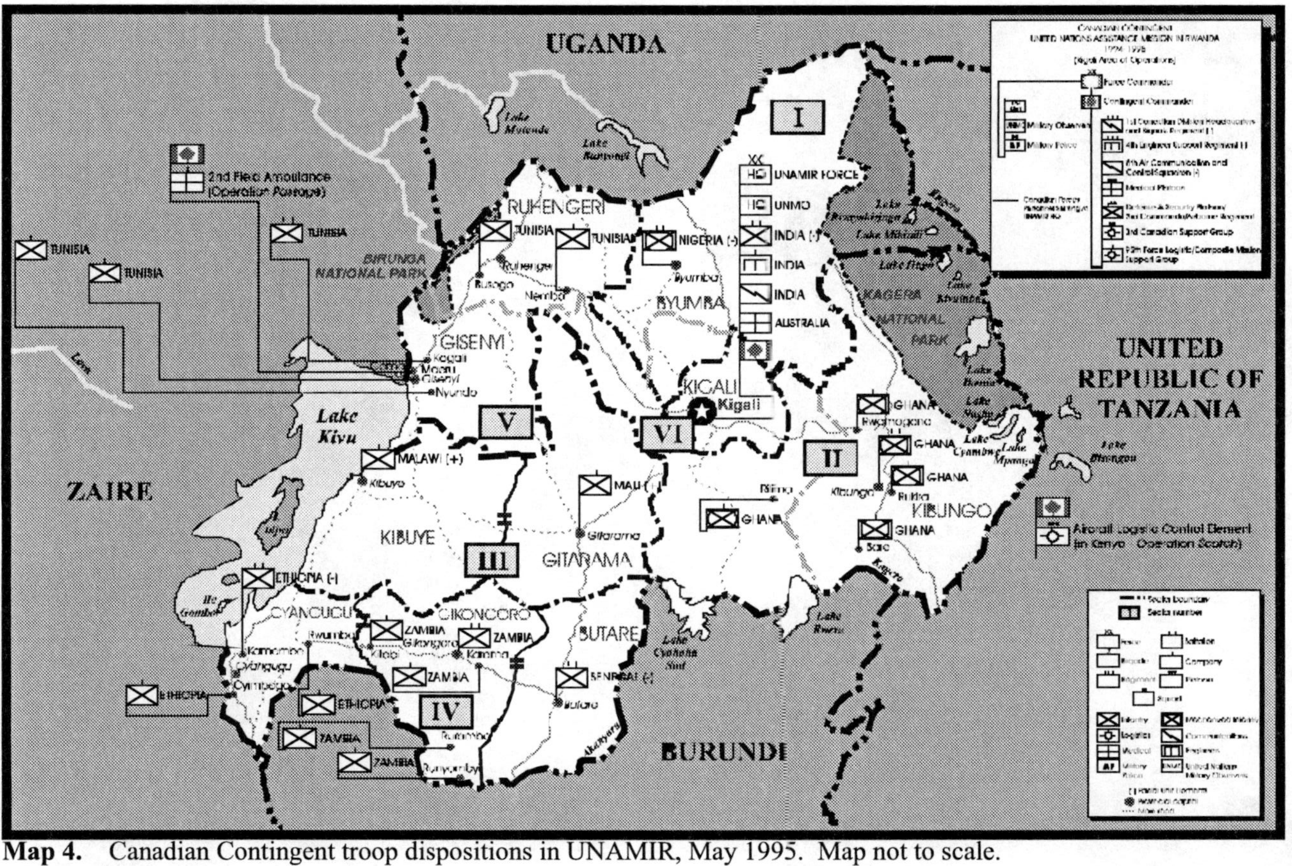

Map 4. Canadian Contingent troop dispositions in UNAMIR, May 1995. Map not to scale.

Figure 8. UN Internally Displaced Persons (IDP) Camp somewhere in Rwanda.
(author's collection)

Figure 9. Canadian patrol greeted by Rwandan children at an IDP camp.
(author's collection)

Figure 10. Rwanda: Canadian medics pause for a photo during a humanitarian visit.   (author's collection)

Figure 11. Marareu, Rwanda 1994: two brothers greet a Canadian soldier.
(author's collection)

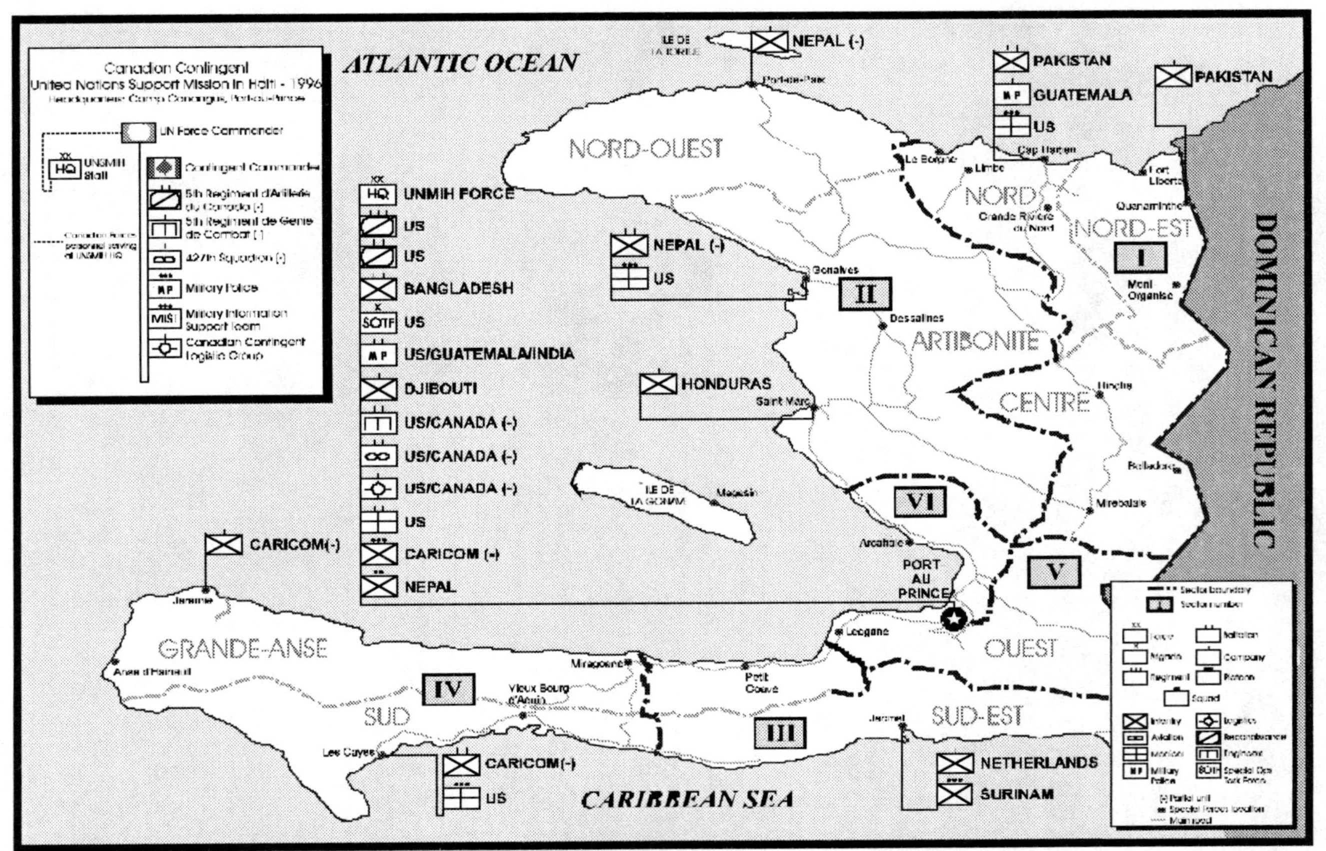

Map 5. Canadian Contingent troop dispositions in UNMIH, November 1995. Map not to scale.

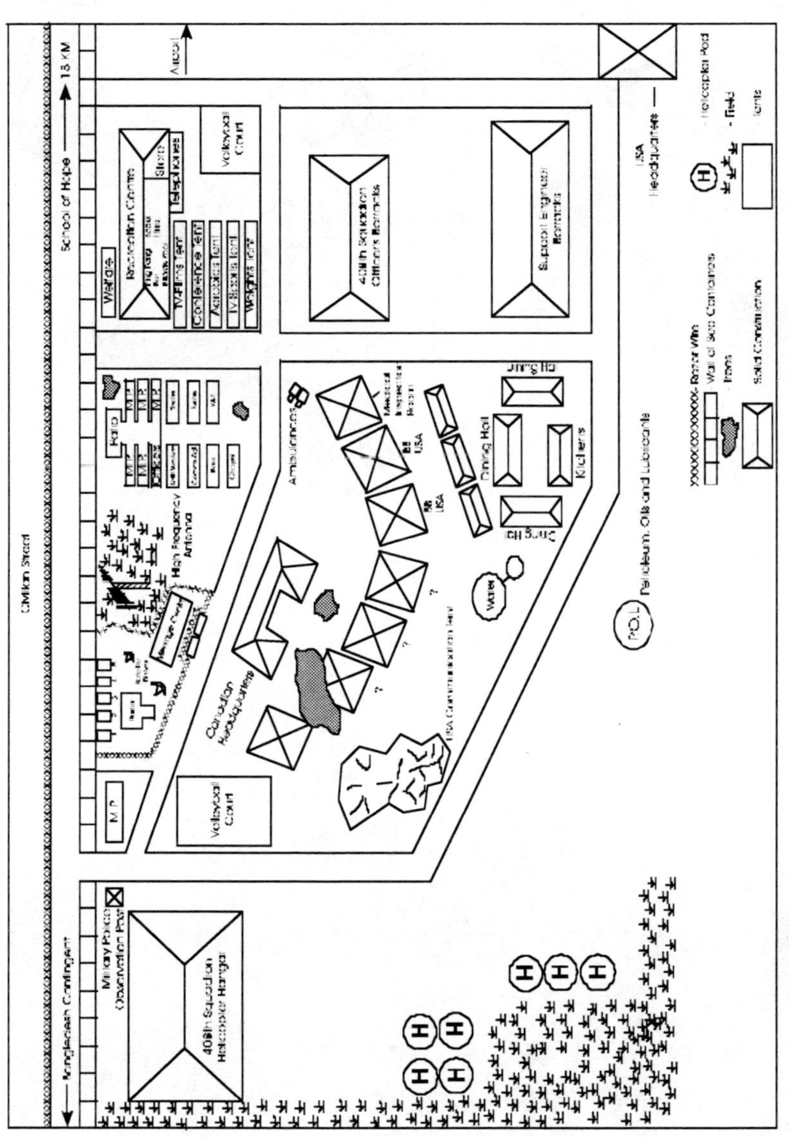

Map 6. Camp Canangus, Port-au-Prince, Haiti. Map is based on a sketch provided to the author by a soldier serving with the Canadian Contingent assigned to UNMIH. Map not to scale.

Figure 12.   Haiti: a Canadian soldier distributing aid to an orphanage.
(DND photo)

Figure 13.   Camp life at Camp Canangus, Port-au-Prince: washing a helicopter and sleeping cots.
(DND photo)

Figure 14. A Canadian Forces electrician helps wire a pump near the town of Gelee, Haiti.

(DND photo)

Figure 15. A Canadian soldier teaches English at a UN-run school in Haiti.

(DND photo)

Map 7. Canadian cities and Canadian Forces Bases mentioned in the text. Map not to scale.

Rwanda

The Canadian Contribution to Rwanda

Mission Background

The United Nations Observer Mission Uganda-Rwanda (UNOMUR) was created in June 1993, following ongoing efforts to resolve the conflict between the Government of Rwanda and the Rwandan Patriotic Front. UNMOUR was based in Uganda and monitored the Uganda-Rwanda border to ensure that no military assistance reached Rwanda by way of Uganda. However, efforts to reduce the number of arms entering the country did little to abate human suffering in Rwanda. Accordingly, the UN Security Council adopted Resolution 872 on October 5, 1993 which authorized the establishment of the United Nations Assistance Mission for Rwanda (UNAMIR). That same resolution also called for the administrative integration of UNOMUR and UNAMIR.

UNAMIR's mandate had several facets. The mandate's main objective was to make the city of Kigali a weapons secure area and to assist in establishing a secure environment until the final period of the transitional government leading up to democratic elections. It was tasked with mine-clearance operations and to investigate, at the request of the parties or at its own initiative, instances of alleged non-compliance with the Arusha Peace Agreement. Also included was the monitoring of the repatriation process of Rwandan refugees, the resettlement of internally displaced persons (IDPs) and assisting in the coordination of humanitarian activities.

Following the wide-spread violence precipitated by the death of the Rwandan president Juvénal Habyarimana under mysterious circumstances in July 1994, the UN evacuated most of the 2,500 personnel assigned to UNAMIR. A small 450-strong garrison remained in Kigali to attempt to negotiate a cease-fire and to coordinate humanitarian assistance. Within this small force, the Canadian presence during the war included the Force Commander, Major-General Roméo Dallaire and 10 CF personnel, including military observers and headquarters staff.

Expansion of the Canadian Contingent

In light of the human catastrophe in Rwanda, Canada increased its contribution in order to provide further assistance to the Rwandan refugees. This additional commitment, Operation *Scotch* and Operation *Passage* concentrated in three essential areas: humanitarian airbridge; medical services; and the purification of water. The initial CF troop component consisted of an:

- Aircraft Logistic Control Element (ALCE), Nairobi, Kenya — Operation *Scotch*

  With an authorized strength of 45 personnel, the ALCE originally deployed to Nairobi in April 1994 with two CC-130 Hercules transport aircraft to deliver humanitarian assistance as well as transporting over 6,315 passengers in and out of Kigali. In addition to this service, Canada assisted in the extraction of 600 Belgian paratroopers flown in to assist in the civilian withdrawal. Later, these aircraft also participated in the withdrawal of 500 Belgium troops and 700 Bangladeshi UN troops during the downsizing of UNAMIR.

  For almost four months, Canada provided the only cargo aircraft to lift UN supplies into Kigali. With one aircraft redeployed to Italy to support Operation *Airbridge* in the former Yugoslavia, the CF dedicated the other CC-130 to the UN until its redeployment to Canada at the end of August 1994.

  During this period, Canada also provided the UNHCR with a CC-130 to transport humanitarian supplies to Nairobi to Kigali and Goma in Zaire (now known as the Congo). This service ended in September 1994 after delivering over 2,580 metric tonnes of relief supplies.

- 2nd Field Ambulance, Maeru, Rwanda — Operation *Passage*

  With an authorized strength of 247 personnel, 2nd Field Ambulance was put at the disposal of the UNHCR to help care for refugees. It was established in Maeru, Rwanda near Goma, on the Zaire border.

  This autonomous Canadian unit was composed of approximately 110 medical personnel, a platoon of 20 to 30 security troops, 25 to 30 support staff, 15 to 20 combat engineers and a 15-member headquarters team.

  Less than a month after the 2nd Field Ambulance arrived in-theatre in July 1994, it was declared operational. Before the 2nd Field Ambulance redeployed to Canada in October 1994, its personnel screened 22,056 inpatients with an average bed day of nearly 2,000 patients.

The Canadian Contingent in Rwanda

On May 17, 1994, the UN Security Council under Resolution 918, authorized an expanded UNAMIR force of 5,500 personnel with the mandate to contribute to the security and protection of displaced persons, refugees and civilians at risk in Rwanda; and to provide security and support for the distribution of relief supplies and humanitarian operations. The mission was authorized to take action in self-defence against persons or groups who threatened protected sites, UN personnel or the means of delivery of humanitarian relief.

The 430-person Canadian Contingent in UNAMIR included:

- UNAMIR Headquarters

  Canada provided a total strength of six personnel to the Force Commander including three NCO's for clerical and transportation support, the deputy chief of staff, a mission advisor to the Force Commander and a PAO. Although the initial Force Commander Major-General Romeo Dallaire was Canadian, technically, he was employed directly by the UN and was not a member of the Canadian Contingent.

- UN Military Observers (UNMOs)

  Canada provided 20 military observers to various sectors of Rwanda to monitor the peace process and serve as the eyes and ears of the Force Commander.

- UN Military Police (MP)

  In late November 1994, Canada deployed (for one year) eight military police as part of the newly formed UNAMIR MP company of 70 personnel.

- 1st Canadian Division Headquarters and Signals Regiment (1st CDHSR)

  Deploying to Rwanda from CFB Kingston, 260 CF signal personnel, drawn mostly from 1st CDHSR, played a key role in providing communication support. To carry out its task, the regiment deployed elements of its signals squadron and support squadron. The 148-strong signals squadron was responsible for ensuring operational communications among all in-theatre units and UNAMIR HQ. The 112-strong support squadron was responsible for administration, vehicle maintenance, medical and logistical support for the regiment.

- 4th Engineer Support Regiment (4th ESR)

  Deploying to Rwanda from CFB Gagetown, 30 personnel drawn from the 4th ESR assisted in the building of infrastructure necessary for a six-month tour. In addition to water purification, carpentry, electrical work and plumbing, engineers voluntarily conducted an extensive Explosive Ordnance Disposal (EOD) operation in the local Kigali area. Lasting over a month, the EOD team disposed of more than 450 pieces of unexploded munitions. In September 1994, the majority of the engineer team returned to Gagetown, leaving behind a small group to maintain the contingent's infrastructure and to operate two Canadian-designed and built Reverse Osmosis Water Purification Units (ROWPU) and water holding equipment. Each ROWPU was capable of producing 50,000 litres of drinking water per day.

- 8th Air Communication and Control Squadron (8th ACCS)

Drawn from CFB Trenton and various Air Command bases, a group of 24 airfield support personnel including: air traffic controllers, radio technicians, radar technicians and other related trades deployed to Rwanda for 60 days. Less than 24 hours after their arrival on July 30, 1994, the 8th ACCS had Kigali airport operational and officially controlled the landing of the first plane. During the first month of operation, the 8th ACCS handled a constant humanitarian airflow of up to 66 aircraft per day.

During August and September 1994, the 8th ACCS and a small US Air Force contingent handled the equivalent of 10 years of normal Rwandan airport traffic. Additionally, the detachment refurbished the airport infrastructure with the repair of airport landing lights, transmitters and generators damaged during the war. At the end of August, the Canadian airfield support staff began a training program for local staff in various airfield support functions. Labeled a success, this training programme allowed the local staff to assume full responsibility for the airport's operation and allow the 8th ACCS to return to Canada at the end of September 1994.

- Medical Platoon

To ensure adequate medical treatment of Canadian personnel, a medical platoon of 27 personnel deployed to Rwanda to support the contingent. In addition to their responsibilities of providing preventive and tropical medical support to CF personnel, they treated more than 250 patients other than Canadian military staff. The platoon conducted five Air Medical Evacuations and assisted the local authorities at various medical facilities in Rwanda.

- Defense and Security Platoon

To provide security for CF personnel, a 37-person platoon from the 3 Commando, the Canadian Airborne Regiment deployed to Kigali, Rwanda from CFB Petawawa. Its primary responsibility was to ensure the security of the 1st CDHSR. However, the platoon did participate for two weeks in routine security patrols in the southeast sector of Rwanda until relieved by Ghanaian units.

- 3rd Canadian Support Group (3rd CSG)

Deploying to Rwanda from CFB Montreal, the 37-person 3rd CSG not only provided logistical support to Canadian personnel; but, it was the only support organization during the initial deployment of the extended UNAMIR force. Since its arrival in mid-July 1994, the unit consolidated UN-owned stock which had been found throughout the capital of Kigali. During its deployment in Rwanda, the 3rd CSG had mounted more than 120 convoys during which they handled and delivered more than 500 tonnes of fuel and more than 17,000 tonnes of freight and material.

Humanitarian Initiatives

Throughout the Canadian Contingent's deployment in Rwanda, members of the various unit had gone out of their way to assist the UN humanitarian effort. Days after setting foot on Rwandan soil, Canadian soldiers started working to assist the government in rebuilding the country.

The signal regiment supported an orphanage of 600 near Gitarama, southwest of Kigali until it departed Rwanda in February 1996. When the orphanage was found, the conditions were deplorable. The children were sick, had no clothing, mattresses or blankets. Immediately, unit personnel volunteered their off-duty time to improve the orphans living conditions. CF personnel purified water, offered medical treatment and support to the orphans. Technicians from the regiment repaired Rwanda's only international communication satellite dish which had been seriously damaged during the war by mortar and small arms fire. In collaboration with the local government and an Irish NGO, it was decided to move the orphanage to Gitarama in late August 1994, due to the lack of proper living quarters. CF personnel cleaned, fumigated and painted the new building before moving the children in September 1994.

In an attempt to inform the Rwandan refugees the war was over and that they could now return home, members of the unit designed, produced and air dropped 300,000 pamphlets over the refugee camps in Goma, Zaire in August 1994. In November 1994, the situation in Rwanda improved and the Canadian Contingent assisted in moving displaced persons from southwest sectors to different villages in Rwanda. This move was known as Operation *Homeward*. From the period of November 1994 to December 1995, CF personnel assisted in the relocation of more than 3,000 IDPs. Operation *Retour* replaced Operation *Homecoming* at the end of December 1994, the objective of which was to coordinate all UNAMIR, NGO and Rwandan government personnel and assets in returning an estimated 19,000 IDPs.

In late 1994, the CF contribution to UNAMIR was re-evaluated by request from the UN, and in late January 1995, a new organization, the 95th Force Logistic Support Group (FLSG) deployed to Rwanda with an authorized strength of 85 personnel. In cooperation with a UN civilian supply contractor, the 95th FLSG developed an integrated supply system to improve manpower management, maintenance and the distribution and acquisition process. In July 1995, the unit was replaced by the 95th Composite Mission Support Group (95th CMSG), also consisting of 85 personnel.

Key Events involving the CF in Rwanda:

- In December 1995, the UN Security Council revised and extended the mandate of UNAMIR. The revised mandate reduced UNAMIR strength below viable levels so the Government of Canada did not support this modified mandate.

Therefore, it was decided that the 95th CMSG would not be replaced at the end of its six-month tour.

- On December 24, 1994, Corporal S.F. Smith of the Canadian Airborne Regiment died in the line of duty while serving in the field with UNAMIR.

- The CF contribution to UNAMIR ended in February 1996, when the last Canadian Blue Berets left UNAMIR Headquarters in Kigali.

This section is based on public information documents produced by the Canadian Department of National Defence Office of Public Affairs.

Letters from Rwanda

The following letters detail the correspondence of eight men and women (of varying ranks and occupations) who served in various aspects of the Rwanda mission previously mentioned and who wrote to Jane Snailham from August 1994 to January 1996. Wherever the location of the peacekeeper writing the letter is unknown the location is noted as "somewhere in Rwanda...."

The following letter was written by a 24-year-old corporal who served with a communications unit assigned to the Canadian Contingent of UNAMIR:

Gisenyi, Rwanda
August 13, 1994

Dear Jane,

Hi there! How are things back in the good ole' Maritimes? Great, I'm sure. I've been in Rwanda only two weeks and already I miss home. Not too much, though, I do enjoy my work.

I was surprised and glad to receive your letter. We share a common bond — we both take pride in what our country does, and we both like to show it. My work consists of going far away on challenging operations and yours consists of letting us know that you care. Both are equally important.

This is my third tour, my first, being Iran in 1988, and my second, to the Persian Gulf in 1990. One thing always on my mind is whether the typical Canadian realizes what we are going through in places such as Rwanda and if they appreciate it/us.

For the duration of my tour, Jane, myself and three other Canadians will be posted to Gisenyi, Rwanda. We are in the northwest (Sector Five), right on the border with Goma, Zaire, and the huge refugee camp which you've surely seen on the news. It is a very mountainous area with a couple of volcanoes, one which is a little unstable which causes a beautiful red glow in the night sky.

We are living in a hotel on Lake Kivu [which forms Rwanda's western border] which was the last stand of the government forces before they fled. The building and surrounding area also was the home to hundreds of thousands of refugees stationed one kilometre from the border.

So what are four Canadians doing here? We provide communications to a group of UNMOs from such countries as Russia, Austria, Bangladesh, Uruguay and Nigeria. These UNMOs monitor and coordinate relief efforts for the refugees as they return to Rwanda. Communications is something people home sometimes take for granted. There are no telephones working in this country, so we are the only way of passing info.

Anyhow, to answer some of your questions, I am 24 years old and I've been in the army for seven years. I am from Cape Breton, which is where my parents have retired. Since my father was in the service, we floated between Greenwood, Summerside, Prince Edward Island and Halifax! Yes, I spent Grades 7 and 8 as well as Grade 12 in Darmouth High School and spent my first four years in the army in Halifax. I have a sister, a brother-in-law and a niece as well as numerous cousins, uncles, aunts, etc. Halifax is where I grew up and I call it home.

My common-law wife is stationed at UN Headquarters, Kigali. I see her once every two weeks, if I'm lucky. She is a great person and I miss her smile and warmth. We write often and sometimes call each other on our satellite phone-link. She is fairly safe there and there isn't many terrible sights. I have seen and smelled enough to last a lifetime.

Anyhow Jane, there is much work to be done, so I must sign off for now. Let me tell you that I very much appreciate you writing to me, way over here in Rwanda. Things do become very routine and boring here, so it's nice to receive mail. So I will continue to write to you (sometimes I'm a little lazy) while I'm here. Until next time!

◆　◆　◆

The following letter was written by a medical assistant who served with the 1st Field Ambulance assigned to the Canadian Contingent of UNAMIR:

Kigali, Rwanda
August 29, 1994

Hi Jane,

I'm a soldier writing to you from Rwanda! A friend of mine received a letter from you and I was fortunate enough to read it and figured you deserved a letter back. There are not enough people like you in this world. You are a wonderful lady to take time out to think of your fellow Canadians!

I'm a female soldier from Calgary serving with the 1st Field Ambulance. I have been in the military for four years now and this is my first tour! I am a medical assistant by trade and enjoy every minute of my job. Tours like this, make me very proud of my job. I can honestly say, I love my job!

I can't complain too much. It's not as bad as the media makes it out to be over here in Rwanda. Our support staff have put up showers for us, and some pretty keen looking out houses! We are supposed to get a television and video cassette recorder soon, but who knows! So, all in all, it's not too bad.

We have set up a 100-bed hospital here with an oral rehydration station and a treatment and screening area. This is a definite learning experience here, medically speaking. We are seeing diseases and illnesses we will never see again back in Canada. It is such a thrill to be a part of this mission! The Africans appreciate us being here. It's comical. They have never seen toilet paper before!

One of the biggest differences I notice here compared to back home is the kids. Here, a simple lid off a bucket and a stick will amuse the kids for hours. Back in Canada, the kids need $500 bikes, $100 skateboards, etc., and it may keep them busy for one-half-an-hour to an hour, then they move on to their "Game Boys" and computers. It's a sin to think so much money is wasted on unnecessary toys!

Well Jane, I must go now. It was a pleasure writing to you. It's a warm feeling to know someone back home, other than family cares about us! You are a role model Canadian that I admire! Please keep writing. I would love to stay in touch. Say "hi" to you husband and son for me. God bless you, Jane, and I am thinking of you.

◆   ◆   ◆

The following letter was written by a medic who served with the 3rd Medical Platoon, 2nd Field Ambulance assigned to the Canadian Contingent of UNAMIR:

Matura, Rwanda,
August 29, 1994

Dear Mrs. Snailham,

The letter you sent to a fellow Canadian soldier was quickly passed around the camp. I want you to know how nice it was to here that other Canadians support us. Of course, all of us here know our families are proud of us, but we rarely here from the general public.

I am a medic from the 1st Field Ambulance at CFB Calgary. There are 25 of us from Calgary. As you know, we pulled personnel from bases across Canada for this tour. We got off to a rough start as we had to wait a few days for equipment to arrive, but the hospital has been functional for two weeks and business is very good.

As well as the hospital, we have a medical team that travels each day to take over where the British used to be. We also send medics to pick up refugees along the roadside and bring them to the hospital for treatment. The town we are in is called Matura, Rwanda. We are about 90 kilometres north of Kigali, Rwanda. Our plane landed in Kigali and we spent the night in a stadium (soccer) with the Canadian UN. It took five hours to get here from Kigali. The best part of it is we saw the beautiful countryside of Rwanda. It's very hilly mountains. We have three volcanoes about 10 kilometres. from us. We call it our back yard. One of them glows in the evening and early morning. We can see it very clearly.

I must tell you where we are living. You will have a good laugh! It's a milk factory without the cows. It's a big brick building.

The thing that surprised me about Rwanda is it's cold at night. It reminds me of a Canadian fall. But this is Rwanda's winter. Spring for us is September 21st.

Well, I think I will sign off for now. Thank you again for supporting our mission. I only know a few people from Nova Scotia. A very good friend of mine is posted in Halifax. If I visit her, I certainly will look you up. I would love to hear from you again. Take care.

◆　◆　◆

The following letter was written by a soldier Jane had corresponded with in the former Yugoslavia and was now serving in Rwanda:

Somewhere in Rwanda
August 31, 1994

Dear Jane,

Well, here we go again. Actually, it's quite funny. I finally got summer leave, so I figured I'd go home to Cape Breton, NS to work on the land. Four days into my vacation and the phone rings — six days later I get off the plane in Kigali, the capital of Rwanda. After unloading the plane and a quick briefing from an officer who met us, it was off to the border of Rwanda and Zaire. The trek was quite an eye-opener as it was hard to digest the amount of people of all ages walking through the mountains on the only road to Goma. Four-and-one-half hours later, we arrive at a milk factory which has never been used, so it was an ideal location for the medical personnel to sleep in.

We spent the next week-and-a-half burying dead bodies which littered the compound, corn fields, etc. and clearing the ammunition, old grenades, etc. We were lucky to have a high fence surrounding the compound, so it was only mending broken holes and fixing up a few houses in the area where our platoon is now in. Two weeks later, the medical personnel arrive, but no supplies in the aircraft, which was chartered by the military and refused to land in Rwanda. It landed in Uganda airport which made the hospital getting set up a bit late. Our job, as for the airborne [regiment] is concerned, is to protect the medical personnel as well as do escorts for various people from the UN. We are not wearing the UN Blue Beret as the mission is from the Canadian government. So the maroon beret is again in Africa.

Our first mail day, we received letters from Canada and some of them were addressed for any soldier. As I walked into one of our rooms, in the distance, I could see an envelope and letter hanging on the wall. Well Jane, it was from a distance I saw this and I knew right away it was from you, eventhough, I could not make out the writing. But the Red Maple Leaf, you put on all your letters, gave it away. I started laughing and my friends asked me, "What's so funny?" I said I know this wonderful lady very well and, of course, they said, "Sure you do." This happened about two weeks ago and I just found time to sit and write.

It's quite different from Yugoslavia that it's almost the same. I know that's hard to say, but everyday here brings unbelievable events, that it is very difficult to try and understand. So I've decided this time not even to bother to try and understand and try and laugh at it, but that's hard to do. The children again are suffering along with everyone else. The graveyard we started digging only for patients that die here, is getting over full. On the second day here, I found a hand sticking out of the ground. We figured we better dig it up and move the body as it was close to where the hospital was going — as fear of disease spreading. After 38 bodies, we stopped digging. Ladies with the babies still strapped to their backs were thrown in this pit. It was very hard to swallow.

I think I'll change the subject. I don't know if you heard or not, but my wife had a baby boy. We have a satellite phone and I call home every week to see what's going on. Everyone there is fine. Hope this letter gets to you soon, as I believe I'll be coming home in 30 to 60 days, depending on the government's approval. So please write back if you not already done so. It's good to see your supporting the Canadian troops. God bless you, Jane. You're a real soldier.

◆   ◆   ◆

The following two letters were written by a 26-year-old sergeant who served at UNAMIR Headquarters and is the wife of the 24-year-old corporal who was mentioned earlier on pages 127-128, and who had served with a communications unit assigned to the Canadian Contingent of UNAMIR:

Kigali, Rwanda
October 8, 1994

Dear Jane,

It was great to receive a letter from you! Yes, it is very nice to know that we are not forgotten here in Rwanda. I don't know how much my husband has told you about me, but here goes.

I am 26, and originally from Quebec City. I have been posted to Kingston, [Ontario] and the regiment for four years now, and my boyfriend and I are due for a posting this year. Back home, my main hobby is my horse, Lucky. I really miss her over here. I am used to going out to the barn for about two hours everyday, so I think I'm going into withdrawal symptoms. I am a confirmed horse nut. Lucky and I, usually go to all the horse shows around Kingston, competing in show jumping. Not this summer, maybe next year!

I work in Kigali at the UN Headquarters. Our living quarters are not as nice asmy husbands,s. He is ina five star resort, complete with pool, bar, restaurant, beach, etc. We live in the Amahoropo Stadium which was a sports arena before the war. During the war, it housed approximately 10,000 refugees. It was a very big mess to clean up when we got here, there was garbage, clothes, excrement and dead dogs everywhere. It took us about a week to clean up, but now things are pretty good. The engineers rigged up a pump so we could have showers. We have messes to eat in and a place where they show movies every night.

I live in a small room with five other girls. We all work shift work, so, it is possible to have some privacy! The weather here isn't quite what I expected. It is much cooler at night than it is back home during the summer. During the day, however, it gets very hot and by 0930, it is usually already 30 degrees Celsius. The rainy season has started here. It rains about twice a day. Once in the morning, then about 1600, it rains again for about an hour. Once the rains stops, though, the sun comes out again full force. I was expecting days of non-stop rain, so I was pleasantly surprised. The only bad thing about the rainy season is the bugs! All sorts of weird, creepy crawlies have come out of the woodwork. I absolutely hate spiders, and there are some dandy ones out here.

I have been promoted to sergeant. It is nice to be recognized, however, it also means that now when my husband comes to visit, we cannot eat in the same place anymore, nor can we watch movies together, either. So it has mixed blessings.

Well Jane, thanks again for writing, it means a lot to know someone is thinking about us! Take care!

Kigali, Rwanda
December 14, 1994

Dear Jane,

Thank you very much for your letter and Christmas ornament. I am very sorry that have not written sooner, but my husband and I were gone on our two week leave and since coming back we have been very busy.

It is finally starting to look a bit like Christmas around here. We put up a Christmas tree last weekend and we have been receiving many cards from school kids which we have hung up around the front foyer of the stadium. They also have a small radio which plays tapes of Christmas music throughout the day. It is just enough to remind us that Christmas is indeed here, whether or not we want to accept it. A lot of people would rather forget about Christmas here and just celebrate when we get home. There is a group of performers here today to give a Christmas show. There are 10 of them [as well as] a band with three singers and a comedian. They are all here from Montreal, so their show is 50 per cent French content.

Only six weeks left! Everyone is looking forward to going home. My husband

and I will be some of the last out, as we will be driving our vehicles (to be put on the trains in Kampala, Uganda), then we will fly out of Entebbe Airport. So, we should be home around the first few days of February. Then we are going to take a well deserved and much needed month off. We will be spending a week in Kingston, getting our car out of storage, finalizing claims, getting medicals done, then we will drive down to Cape Breton to see my husband's parents. Along the way, we are going to stop in Quebec City to visit my mother. We are probably going to go to Halifax to visit my sister-in-law for a few days as well. I am very much looking forward to it.

Things here are very quiet. Not much is going on anymore. I went to visit my husband the other day and he took me to meet someone. If you have ever seen the movie *Gorillas in the Mist*, this woman is in that movie. She was very good friends with the character portrayed in the film, and is now a prominent figure here in Rwanda. Anyways, she is a wonderful lady, and her house is just amazing. She has a beautiful garden which she brought all the bulbs and seeds from either the US or the UK. They are just beautiful. You actually forget that you are in Rwanda when you are at her place. She also tells great stories about the war and the filming of the movie. It was one of the highlights of my tour here. So Jane, I would like to wish you a Merry Christmas and a Happy New Year and thank you for your cards, letters and good wishes. I will write again soon.

◆   ◆   ◆

The following letter was written by a lieutenant-colonel who commanded the 95th FLSG assigned to the Canadian Contingent of UNAMIR:

Somewhere in Rwanda
February 15, 1995

Dear Jane,

Thank you very much for you kind letter. We have just arrived here in Rwanda over the past month and are very much enjoying the change from the Canadian winter. It is very comforting to us to receive letters such as yours, and we do receive a fair number of them. It is nice to know that what we do is appreciated by so many and I thank you again for your interest.

We are pretty busy here working seven days a week until about 2000 or 2100. each night. The work is fun, however, as it is a new thing, we are trying to integrate the UN civilians, (well over one hundred), and the primary contractor who consists of 150 Americans, Brits, Aussies, etc. and 1,200 to 1,400 locals into one organization. We have about 85 Canadians trying to make it happen so we can do the inventory control, fleet management and maintenance, resupply of everything from water and rations to spare parts and toilet paper. The system in place is not working well — we are meant to make it better. It should be relatively easy to make it better as it is so bad. I do not think it will be possible to get it right.

Life here is more than a little hectic, but the good thing about working seven days a week you don't have to worry about being bored or getting into trouble. We are still confined to our "compounds" at night and for good reason, but things are visibly better if not necessarily safer every day. We are attempting to settle in, move and plan the stand up of the integrated logistics concept at present.

We have a strong team quality wise if not number wise and we will have no trouble making a mark. I hope it will be a positive and enduring one. There is more than enough work to go around with fleets estimated at 1,000 heavy and 800 light vehicles. I say estimated as no one seems to actually know! There is at present no "fleet management," consistent or reliable spare parts, procurement, inspection or repair schedules, etc.

On the supply side, there are no scales of issue and no real inventory management, let alone controlled stores, lists and the like. Everything is from the seat of the pants, but unfortunately, not mine! The fellows we've dealt with so far are cooperative and friendly, but one would not expect any less as they have been doing both the line (the work) and staff (the planning) functions and we are the answer to their prayers in many respects.

The unit looks really good, especially when one considers it is a composite unit. The guys/gals are great and generally very experienced and easy to work with.

Rwanda is beautiful country with a year round temperature of 25 to 27 degrees Celsius and very little rain other than during the rainy seasons. Being in this area of the world always serves to remind me how good we have it in Canada. I will close this by saying thank you again for your kind letter and thoughts — they are appreciated.

◆ ◆ ◆

The following nine letters were written by a naval lieutenant who served as a UNAMIR PAO:

Somewhere in Rwanda,
February 16, 1995

Dear Jane,

I just wanted to thank you for your letter which found its way to my desk.It is very nice to receive a letter like that, so cheery and kind.

I have been here now for three weeks and there is no doubt that this place takessome getting used to the culture, the climate, the way things get done, the people and of course, all the languages we deal with here in UNAMIR HQ.

Also because of my job as the PAQ, I have the opportunity to get out and aroundthe county quite a bit. This is good and bad because I an afforded the chance to see things first hand, but unfortunately, a lot of what I see—results fromthe war — are often not as cheery as your letter.

However, I can say I am very proud of the work that we are doing here in Rwanda, life is coming back to the country and you can literally see it day by day. Traffic and pedestrians in the city are noticeably increasing and there seems to be more life in the air than when we arrived.

Well, I should get back to work, but thanks once again for your letter, Jane, and I look forward to hearing from you soon.

Kibeho, Rwanda
April 18, 1995

I conducted a media flight by UN helicopter to Kibeho IDP Camp this morning. We arrived at the camp to find that all the IDPs were either already concentrated towards the centre of the camp or walking en masse along the roads to the centre of the camp, surrounding the ZAMBATT [Zambia Battalion] Company HQ compound. We flew over a number of times.

There do not appear to be a huge presence of the RPA [Rwandan Patriotic Army] from the air, those that I saw were merely walking along the roads. There was no sign of mass panic and no signs of any violence, but it was clear that no one remained in their huts. Also, the IDPs have taken off the plastic blue sheeting covering their huts, which gave the impression that the huts had been burned or destroyed, but this in fact was not the case. I observed some RPA talking with some ZAMBATT soldiers. Nonchalant waves from the ground from ZAMBATT gave me the impression that they were not in danger, but I passed messages to UNAMIR HQ through the helicopter radio describing the scene from the air. As there was no ZAMBATT security present at the helipad, the pilot opted not to land and we returned to Kigali.

Kibeho, Rwanda
April 19, 1995

I accompanied [the] media to Kibeho Camp and found that there was now a large presence of RPA soldiers who surrounded the people still concentrated in the centre of the camp. They were not hostile in any way and I remember that at the screening point, they were quite efficiently and calmly registering people and sending them on to the transportation point. I was impressed with how professionally they were dealing with the situation.

Kibeho, Rwanda
April 21, 1995

I accompanied another media visit to Kibeho and now the sanitary conditions (or lack thereof) made the camp quite inhospitable. There was more tension, fear and hunger with the IDP's, but the situation did not seem out of hand. The RPA were not quite as calm as they had been and it was evident that after three days of being crammed together, being forced to leave their huts and go home, having no sanitation and lacking sufficient supplies of food and water that the IDPs were becoming more miserable. However, again there was no overall indication on either the part of the IDP's of the RPA of what was to follow the next day. (They shot 16 IDPs that day.)

Kibeho, Rwanda
April 22, 1995

I arrived at approximately 0745 at Kibeho Camp by UN helicopter with journalists from *Reuters*, *Magnum* and *Die Deit*.

The RPA at the first roadblock, located between the helipad and the ZAMBATT Company HQ, denied the journalists access to not just the camp, but also to the ZAMBATT Company HQ.

As I was in uniform, we agreed that I would go into the camp and find out what was going on and return and brief the journalists later that morning.

At this time, across the hill, we saw a man being chased by two armed RPA down the hill — he was shot at a few times, but not hit and a few minutes later, apprehended.

Throughout the rest of the morning, there were sporadic bursts of gunfire either into or above the crowd, or at specific persons trying to run through the cordon

On entering the camp. I first made my way to Medicins Sans Frontieres [Doctors Without Borders or MSF in French] building, located next to the ZAMBATT Company HQ. On entering the compound, I faced approximately 50 persons with severe, fresh machete wounds to the head, face, neck, back, arms and legs. There appeared to be only two local MSF staff on hand, with no medical supplies and there was not much they could do.

I then passed by the back of the ZAMBATT Company HQ where the Zambians showed me an IDP who had tried to hide in one of their pit latrines and was buried up to the head in excrement and was either dead or unconscious.

By 0825, I had made my way along the road towards the ZAMBATT Company HQ location, found in the centre of the camp. The crowd was very tense — likely from being packed together as they were, but also from four days of little sleep, food and water, horrendous sanitary conditions, night machete attacks and fear of their future, but specifically, I would say, as a result of the sporadic gunfire that was occurring.

At 0830, I witnessed a man trying to run through the cordon past the RPA. He was shot in the back at very close range by an RPA soldier chasing him. I tried, along with two Zambian soldiers, to get over to see if he had survived, but we were prevented from doing so by the RPA.

For the next hour-and-a-quarter, I made my way to the back of the ZAMBATT Company HQ location, but due to the crowds, was unable to gain access to the compound. The crowds were crushing up to the compound and there was a general tenseness and misery about the situation amongst the IDPs.

At some point, I saw another man being shot by the RPA as he tried to run down a hill. Again, we were denied access to his body, but within 15 minutes, he had been buried on the spot by the RPA.

During this time, ZAMBATT soldiers began to move the very frightened crowd back from the ZAMBATT company location. This was a slow, but steady process. People were generally confused, fatigued and despondent. As they passed by us, some would indicate that if they left their throats would be cut, others made half-hearted attempts to walk towards the RPA who pushed them back into the crowd and others were dehydrated enough to drink muddy water from plastic sheets strewn about the ground. Many children had lost their parents and were wandering around aimlessly.

A UN truck had been overtaken by the crowd and men could be seen atop the cab and others trying to get in. It had to stop as it made it's way towards the compound due to the density of the crowd.

After assisting in directing people back, I made my way through the crowd to the ZAMBATT compound. I had been in Kibeho on Wednesday and Thursday that week, but never had the crowd been packed together so tightly. It was essentially a question of forcing our way through the crowd and for the first time in my many visits to Kibeho over that past week and over the past three months, I distinctly felt that there was an air of danger, fear and tension among the IDPs.

As we pushed our way through the crowd towards the entrance of the compound, the crowd cleared, but only because the road was covered in bodies of dead, dying or injured people — to the point where it was impossible not to pass by without at some point stepping on someone. I would estimate that there were about 20 people: men, women and children, laying in front of the compound. However, these were injured not by bullet wounds, but rather machete wounds or having been crushed, suffocated or dehydrated.

I finally gained access to the ZAMBATT compound and over the next two and one quarter hours, we assisted with bringing in injured people — some with machete wounds to the face, having babies/children, both alive and dead, passed to us from the crowd and giving what water we had to the sick and dying. Some of the ZAMBATT soldiers would go out into the crowd and help bring in bodies or injured people. They also were able to clear the pile of bodies at the front entrance, by bringing them into the compound. They also provided a sense of stability to an impossible situation and kept the crowd as calm and organized as they could. However, with the sporadic firing taking place at all points of the camp, as the morning went on, the situation appeared more and more grim. Some healthy men fought their way into the compound, but were apprehended by ZAMBATT and pushed outside the barbed wire perimeter.

At approximately 1035, the ZAMBATT compound received fire to the degree that we all immediately dove behind sandbags for a few minutes. It is impossible to say whether the fire was directed at us, but is certain that it passed all around us — no one was injured in the compound, but could have been in the crowd.

At 1045, I confirmed with an UNREO [United Nations Rwanda Emergency Office] representative who had made his way illegally past the cordon and into the compound, the MSF, UNICEF [United Nations Children's Fund], Save The Children, etc. [aid workers] were all being blocked by the RPA cordon and no medical assistance was getting in.

By this time, there were over 125 people inside the compound to whom we were providing safety, security and sanctuary as well as humanitarian and medical assistance as best we could. I counted 35 dead babies/children that were lined up in the compound. Of the other victims, I would say that the majority were still alive, however, about 15 adults and youths were dead from trampling, suffocation, dehydration, etc. There were also about 75 children sitting in the compound. Throughout, the ZAMBATT troops were providing water to them and those in the crowd and generally maintaining some semblance of control around the compound, in fact, doing a magnificent job given the conditions.

At approximately 1100, the prefect of Gikongoro [Province located in Southeast Rwanda] drove through the crowd and through the compound, accompanied by [the] RPA. He continued on towards the transportation/screening point at the opposite end of the camp.

Shortly thereafter, I watched another man shot who had been walking past the compound down the hill at the back of the ZAMBATT location. Again, we were denied access to his body and he was buried by [the] RPA within five to ten minutes on the hillside in a shallow grave.

Throughout this time, there was sporadic gunfire throughout the camp. Also, I could see that people walking through the screening point towards the transportation points — as they were, they were being beaten severely by [the] RPA with long, heavy sticks and rifle butts.

In the opposite direction, in a clearing on the hill, a woman was beaten to the ground by three RPA [soldiers] with sticks, then chased and beaten back up into the crowd. If you are wondering why we couldn't do anything, these occurrences were taking place hundreds of yards and tens of thousands of people away, and our hands were full taking care of the sick and dying around the compound and also providing security to the compound. At the same time, ZAMBATT soldiers would venture into the crowd to assist those that they could.

More people with machete wounds stumbled into the compound area and we had them sit in the shade of some UN vehicles in the area. By this time, the women and children were hiding under the vehicles inside the compound.

At 1150, more shots were heard around the camp, this time a more serious and intense volley — the crowd was bordering on panic.

At 1155, a severe rainstorm approached the camp and by 1200, the rain started coming down hard on the crowd. The ZAMBATT soldiers held their positions at the barbed wire perimeter of the compound shouting to the crowd to stay where they were and calm down — others helped move the 35 to 40 dead babies/children into the building for dignity from the rain.

As the rain beat down and the crowd shifted for shelter, heavy gunfire erupted from all over the camp. Within one minute, despite the best efforts of ZAMBATT soldiers to prevent it, the crowd poured over the barbed wire and overran the outside part of the compound, which included the jeep I was in. They did not make it into the walled part of the compound.

Our vehicle instantly disappeared under the crowd. We couldn't see outside the windows or the windshield due to the people crushed against the car, on the hood, on the windshield, on the roof and under the vehicle. For the next one-and-a-half hours, we remained in the vehicle, not only physically unable to open a door for the crush of people, but when I started to roll down my window for air, people tried to force their way inside.

The firing continued from 1200 to about 1250, relatively continuously, then died down to sporadic fire. The people on the roof had broken our antennae so we could not send messages out about our situation, but as we were about 20 yards from a UN truck that had three ZAMBATT soldiers on the cab keeping people from overtaking it, we remained inside the vehicle rather than risking the impossible of forcing our way out of the vehicle and through the crowd.

At about 1300, someone told us through the window that people were being macheted in the crowd behind the vehicle — one minute later, a man's face appeared in one of the windows, split in half with a machete.

Throughout, people were passing babies and children above their heads towards the ZAMBATT compound, which may explain the 250 abandoned children found there later.

At about 1330, the crowd was still crushed together, but two ZAMBATT soldiers forced their way to the vehicle and using sticks, were able to clear a path for us towards the end of the camp. As they did so, they had to pick up and remove bodies from before us, and as we passed, the crowd swallowed up any space that had been provided.

By 1345, we were at the transportation point outside the perimeter of the camp. Sporadic firing could still be heard. After waiting for instructions for about [an] hour, then we were told to go to Butare [the district capital of the province of Butare located in Southeast Rwanda].

On the road to Butare, we measured a 13-kilometre steady stream of IDPs making their way along the road. They were being beaten by the RPA with sticks, were being stopped and having what little possessions they had with them taken, and were being forced to run down the road by the RPA chasing them. On more than one occasion I witnessed local civilians along the sides of the road beating the IDPs as they passed by.

At this point, I stopped being a witness to the events, but am told that at approximately 1730, heavy firing into the crowd took place for an extended period, including machine gun firing, grenades and, I am told, mortar rounds.

Kibeho, Rwanda
April 23, 1995

This morning, the RPA were burying bodies by 0500, in pit latrines and shallow graves. I arrived in the afternoon with another media team and filmed the burying of bodies by ZAMBATT soldiers and also the beginnings of the standoff in the compound. [the] RPA were no longer burying bodies. There were approximately 15 bodies, men, women and children laying in front of the standoff compound. Journalists filmed one "body" being picked up on a stretcher to be buried at a mass grave, only to find that he was feigning death because he immediately jumped up and ran into the compound. We entered the compound and found that estimates of the numbers of people seen in the compound by the journalists varied from 300 to 2,000 which helps show how difficult an accurate count is, given the debris, the children and the setup of the compound.

ZAMBATT troops entered the compound and tried to get people to leave by speaking to them with loudspeakers. Many did so, but the majority remained. Another escaping IDP was shot dead by two RPA and his body was left in the valley at the base of one of the hills of the camp.

Kibeho, Rwanda
April 24-25, 1995

With two other media trips, the sanitary/health situation was getting much worse. Children were coughing and people mostly sat around dejected. On entering the camp, the stench of excrement was becoming terrible and the overall situation was very poor. However, when we spoke to people and asked them if they wanted to leave they replied that they would rather die there than return to their home communes where they were certain that they would be murdered or venture outside the compound as they thought they would be killed by the RPA. This despite the large UNAMIR military observers/ZAMBATT presence and assurances that they would be transported to Butare in UNAMIR vehicles under UNAMIR escort. To this, some of the men replied with jeers and exclamations of disbelief. Still, some people did walk out of the camp. UNAMIR personnel entered the camp and took out bodies of those who died during the night, either from disease or by machete. UN agency representatives were present and the RPA had called many deadlines which passed uneventfully. On the outside, the RPA were simply getting bored and the IDPs were getting more depressed and tired. Water was collected in buckets and cans when it rained. Some women were singing religious songs as we left.

Kibeho, Rwanda
April 28, 1995

Conditions had deteriorated a great deal, to the point where you couldn't go inside the compound area without stepping in excrement. The stench was overbearing, flies were everywhere, and although I am not a medical doctor, it was clear that this was fast becoming a serious health hazard area. Most people were sitting down, but a few men were busy cutting firewood with their machetes. Again, we went in and spoke with the remaining IDPs who again insisted that if they left, they would be killed. Most of the women were despondent and the children coughing and sick. The RPA were bored, but spoke with some of the IDPs in my presence, but not in a hostile manner. In a word, the situation could be described as pathetic. I spoke with many journalists that evening who have been in Africa for many years covering the Sudan, Ethiopia and Goma, [Zaire]. They said that it equaled the worst of what they had ever seen and some claimed it was the worst conditions they had ever seen. The IDPs are literally living in garbage and human waste, but they either will not move out or have been intimidated to stay.

Somewhere in Rwanda
May 17, 1995

Dear Jane,

Things have settled down here somewhat following the Kibeho Massacre — now that was truly a life-experience. I never saw my interview, but I hope it was okay and that people understood the horror of the situation and also just how much good the Zambian troops did to protect lives — it's a tragedy that so many innocent people had to die and I hope I never see the likes of that again. I was glad I was there though as I was the only person in the middle of it all who had a camera and a video camera, so I was able to record a lot of the atrocities, that then served as key evidence in the International Commission into the Kibeho incident. The commission viewed all my video footage and looked at all my photos before concluding that the RPA soldiers had used "excessive force in arbitrarily depriving the IDPs of life and their human rights."

I assume by my piece in the paper, you are referring to a letter to the editor about our peacekeepers helping out at orphanages. They really did a great job and made us all proud, so I sent a letter to the editors of [all] papers across Canada. It was the "letter of the day" in the *Toronto Star, Ottawa Citizen*, and the *Kingston Standard*.

I am now writing hometown articles on all our Canadians serving here, so these should start appearing soon.

◆   ◆   ◆

The following letters are three of four written by a sergeant who served with a support unit assigned to the Canadian Contingent of UNAMIR:

Kigali, Rwanda
August 30, 1995

Hello Jane,

Your letter was circulated amongst the Canadian Contingent. I just received it and thought it very encouraging for a peacekeeper, so far away from his home. It's good to know that people care for the work that their CF are doing. I'd like to be your Canadian pen-friend. It's always good to get letters from home. It keeps my morale high.

Let me tell you something about myself. I live in the town of Oromocto,NB with my wife of almost 17 happy years. We have two boys, one 15 and the other 13. My oldest boy has just started high school, and my youngest son is now in Grade 8. My wife works and is a good mother and a good wife. We attend the Oromocto Baptist Church. I'm a deacon of the church and my wife and I are also youth leaders. We have a good group of boys and girls in our youth group.

I've been in the military for 17 years and am a sergeant. On July 1st, I will have completed my 18th year. Our first posting was Petawawa, where we lived in a PMQ for five-and-a-half years. In 1984, we were posted to Lahr, Germany. There we spent six good years. We saw a lot of country while there. In 1990, we were posted to CFB Gagetown. My first three years, I was with the 2nd [Battalion, the] RCR. During that time, we were involved with Cyprus and Yugoslavia. So as you can see, I was busy. Now I'm with base supply, since that is my trade (supply technician). Now I'm away from my family, once again, in Rwanda. For me, it is hard to be away from my family since we are so close. We do a lot of family orientated things. My wife is not just my wife, but my best friend.

I was brought up in Dartmouth. I lived there for 21 years. My parents moved to Moncton, NB but I stayed behind and worked for my uncle. Then I went to Moncton to attend Atlantic Baptist College. That is where I met my wife. Now, I'm serving in the CF.

It is a different life over here. As Canadians, I think we have a tendency to take things for granted. We have so much and these people have so little. The amazing thing is that they seem to be happy with what they have. For this is all they know. They live in mud houses, and some of them live in straw homes. A lot of them are just one room with about four people living there. My job over here is to order spare parts for the vehicles that are broken down. Let me tell you one thing, though, it's not like Canada. It takes at least three months to get any parts over here. The Canadians have a saying, "Take a Rwanda minute." The reason we say this is that they are in no hurry. A Canadian hour is a Rwanda minute.

Don't get me wrong. This country is beautiful. I've had an opportunity to drive and fly over the rain forests. It is so massive and dense. When you leave the city and go to the outskirts, it looks like time has stood still. They do a lot of manual labour. The people walk everywhere they go. They will walk half a day to get to market and another half a day to get back.

That's a little about myself and my job. It would be good to hear from you again. Thanks again for your letter.

———————— ◆ ————————

Kigali, Rwanda
September 25, 1995

Hi Jane,

Let me tell you what I've been up to since the last time I wrote to you. I'm very involved in the orphanages. The one I visit most of all is Jesus Alive. In fact, tomorrow, I'm going down with five Norwegian nurses to do a scabies treatment. We are delivering 300 new blankets, a wheel barrow, a big cooking pot and utensils.

The city is still growing. We employ at least 200 locals around the camp. I don't know what will happen to them when we leave.

I went on a safari to Uganda at Queen Elizabeth Park. Africa is a beautiful continent. Lots of mountains and an awful lot of bananas. It was good to get away from the camps for a little R&R.

Now I'm back into the old grind. It is hard to be separated from my family. But I keep busy, so the time will go by faster and I'll be home with my family quicker. I go home for two weeks in December, but I'm back in Africa just before Christmas. So my family and I are going to celebrate Christmas before I come back.

This was just a short note to keep in touch. If I ever get to Halifax, it would be nice to look you up. If that's alright with you.

Take care and I hope this letter finds you in good health.

Kigali, Rwanda
October 10, 1995

Dear Jane,

As I told you in my last letter, when you are away from home. it is such a morale booster to get letters from home. It is also nice to know that Canadians do care on the job we are doing on any UN mission.

I keep very busy here. That way the times goes by quicker. This country is coming alive. Everyday I drive to work, there are more and more people on the streets. We are making a difference here. You can see it. More and more business are opening up, more cars on the road and people are getting electricity in their homes.

Last night, we had a mess dinner to celebrate Thanksgiving. It sure didn't feel like it, though. You sure loose track of time and what season it is when you are in a tropical place like this. We are still wearing T-shirts and shorts in the evening. Talking to my wife, she tells me that she has to put the heat on in the morning to go get the chill out of the house. It's hard to picture that.

Well, Block Three leave has just left today. I'm on the last one, which is Block Six. So, as you can see, I have only two more to go before I get to come home. The only problem with that is that I have to return just before Christmas. So, we are going to celebrate Christmas before I come back. Then, when I get back to Rwanda, I should only have a month left of the tour.

We are not quite sure how long we will have left. We have heard that we are the last Canadians that are coming over here. This mission is suppose to close down by December 9th. But because I am in the supply world, my job is to make sure that all the material that the UN contingent have signed from us has to be returned and packed in shipping containers. The best thing about this, is that we should only have at least 800 people to worry about. Getting things back from 800 people is a lot easier than if we had 3,000 people. We are all hoping that we are still out of here on the same date we are suppose to be, which is January 26, 1996.

Well Jane, that's all I have to say for now. I enjoy hearing from you and I have to thank you for your prayers and concern for all us peacekeepers, for I am a very religious man. Thanks again, bye for now.

The following letter was written by a sergeant who served with a logistics unit assigned to the Canadian Contingent of UNAMIR:

Somewhere in Rwanda
November 4, 1995

Hi Maritimer,

I am sitting here in Rwanda, Africa, reading your letter and feeling pretty good. Especially when typical Canadians can realize that the Canadian soldier is serving our country worldwide and making a big difference.

Well, I am a sergeant. My nickname is Q. I am working out of CFB Halifax, as a mechanic in Willow Park. Here in Rwanda, I am in charge the recovery of all UNAMIR vehicles. Also, I am very much involved in the orphanage committee, which takes up many hours, helping out the 20 different orphanages. I have been here since July 16th. Hopefully, the tour will be finished on January 25, 1996.

All of the maintenance crews, have a large hand in the orphanage working groups. This brings a great feeling of satisfaction and appreciation to our crew, when one of many, many tasks brings a smile to a kid's face.

Canadian peacekeepers, along with other countries in UNAMIR, assist with the building of bridges, supplying food and water across the country, providing medical care, conducting military observer patrols, repairing roads and monitoring the peace process and assisting in humanitarian relief operations.

One thing that bothers me is that there are so many kids without parents, due to the genocide that took place. I have been on a lot of taskings, dealing with the people of Rwanda, and their value on life is completely opposite to what we are accustomed to.

After we complete some of these difficult jobs, we are always reminded that as soldiers we are visitors/guests and not here, to change their lifestyle. But as a Canadian soldier, I feel good if we do something to relieve the human suffering in this strife-torn country.

My wife and girls live in Nova Scotia. It was very hard on her, for me coming here, plus we have built a new house, with lots of work to be done. Also, it is going to be harder on her for Christmas, when I am here in Africa.

Well, that's enough for now, again, it was nice to read your letter, thanks for the concern. Until next time.   ◆   ◆   ◆

The following letter is the last of four written by the sergeant who was mentioned earlier on pages 142-144, and who served with a support unit assigned to the

Canadian Contingent of UNAMIR:
Kigali, Rwanda
November 7, 1995

Hi Jane,

I'm so glad to hear that Quebec is still part of Canada. It was a very close vote. I didn't think it would be that close. I'm with you, Jane, I hope that's it for now.

Now let's see what has been happening since I last wrote you. It's been very busy here. The worst thing is that we don't know what is going to happen to this mission. It is leaning towards a close out. If that happens, we might be home by the end of February. So, I'm doing a lot of paper work, revising requisitions that I've ordered spare parts on.

The orphanage is still going good. In fact, the nurses from Norway have been coming with me and treating the children that have skin problems. We've been doing this for the past in month-and-a-half. You should see the children now. One hundred per cent improvement! It's great to see.

I don't know if I told you in my last letter, but I went to see a shrine a few weeks ago. They have left it the way they found it after the war. It was very sad to see and hear about what torture they went through.

I took a few pictures. I'll have to bring them with me when I come to Halifax, so you can see how beautiful this country is. More and more people are coming into Kigali. The city is getting back on its feet. It's less than a month and I'll be out of Kigali.

Well Jane, that's about it for now. Take care and once again, thanks for your prayers and thoughts.

◆   ◆   ◆

The following two letters were written by a warrant officer who served with the Canadian Contingent assigned to UNAMIR:

Somewhere in Rwanda
December 21, 1995

Dear Jane,

Letters and cards from all across Canada, addressed to Canadian peacekeepers in Rwanda have been flooding into our contingent mail room for the past couple of weeks. When we can find the time, soldiers have been happily responding to any which included a return address. Since I am from the Halifax area, I would like to say hello from Rwanda and take this opportunity to tell you about some of the things we do here.

I am a warrant officer and I have been in Rwanda since July. It is a very small country, being only half the size of Nova Scotia. There are a little more than 100 CF personnel serving with the UN in Rwanda. UNAMIR was established in October 1993 to police a cease-fire between the Hutu and Tutsi tribes who were fighting for control of the country.   UNAMIR is helping the Rwandan people rebuild their nation by assisting in national reconciliation and providing for the voluntary and safe return of refugees to their homes. The job of Canadian peacekeepers is firstly, to provide logistical support to UNAMIR, and secondly, to support humanitarian activities throughout the country, in general.

I arrived in Rwanda during the dry season. It was very hot, about 30 degrees Celsius everyday. The sky was typically a hazy blue and generally, always cloudless. There was no rain for about two months. Two months without rain or fog for an atypical bluenoser is a very unique experience. Initially, I had to drink lots and lots of water to stay hydrated. I am acclimatized now and do not require as much water, but whenever the temperature dips below 20 degrees Celsius I feel chilly. Now, it is the "mini" rainy season and the weather is somewhat more wet and a little cooler. It does not rain for long, but it rains very hard. Sometimes if you are driving, you must pull of the road, because visibility becomes too poor to see properly. Even a 15-minute rainfall can be enough to cause mud slides and flooding. Since we are close to the equator, there is only 12 hours of light, everyday, all year round. The sun rises at 0600 in the morning, and sets by six in the evening. Of course, one thing I don't have to worry about, is how much snow I will need to shovel to get out of my driveway.

Last year, fighting erupted once again. This time it was genocidal in nature, and hundreds of thousands of people died over a period of a few months. It has to rank among the most socially destructive and tragic events of this generation. Men, women and children killed and were killed. Even now, a year later, the evidence of this horrific event still lingers. From the despair of genocide, UNAMIR has assisted the Rwandan people in many ways to alleviate their grim predicament. We have rebuilt and supported schools and orphanages for tens of thousands of children, helping them to deal with the trauma of war. UNAMIR has contributed to the building of extensions to prisons to ease the appallingly overcrowded conditions for those incarcerated. In Rwanda today, over 55,000 prisoners are housed in facilities designed for a maximum of 13,000. Other assistance includes training of police, provision of more advanced medical care and the restoration of national medical services, repairing of roads and bridges, restoration of telephone services and the building of refugee transit camps. These contributions have helped bring back some normalcy to a country severely disrupted by war. Not that long ago, the capital city of Kigali was dead in every sense of the word. Bodies still lay in the street, packs of dogs fattened from the corpses ruled the city, houses were destroyed and there was no electricity, water, nothing. All that seemed to remain was the stench of genocide and children abandoned by war, pathetically wandering the streets, traumatized by the death and destruction they had witnessed. Today, with UNAMIR's assistance, Kigali is essentially, a fully-functioning city. You won't see any dogs, though, since many had to be shot to prevent the spread of disease.

There are over 12,000 orphans in Rwanda. The Canadian Contingent sponsors a number of orphanages and visits them regularly, providing medical care, water supplies, repairs to electrical and plumbing problems, building playgrounds and donating medical supplies, blankets and shoes. It is important to let these children know that someone cares about them and that they have a reason to hope and believe in the future. I helped deliver some food to an orphanage, recently.   The children seemed rather amused at my white skin, but were very happy to see us. Before we left, they gathered outside the orphanage and sang and danced for us. On a separate occasion, I viewed some drawings made by children from another orphanage. Sadly, many of the drawing depicted the terrible experiences endured by the littlest victims.

As we carry out our duties, we travel all around the country. There are only a few good roads, and it can take a long time to travel a short distance. When it rains, the dirt roads become very muddy and slippery, as if covered with a layer of snow and ice. There are few trees, because they have been cut down to clear land for farming and for heating. In many areas, there are buried mines, so we are careful to exercise mine awareness, wherever we go. Everybody has so far stayed safe, but some of us have been very, very lucky. Unfortunately, many Rwandans are still being hurt by these mines. These weapons do not discriminate. Men, women and children are being killed or maimed almost everyday. There is still an ongoing low-level insurgency war which can make traveling through some parts of the country somewhat entertaining on occasion. Certainly, we try to avoid moving by night wherever we are and have adopted measures to enhance our security when we do.

Rwanda is among the poorest countries in the world and life here is pretty basic. Most people are farmers and they live in very small houses made of clay or mud bricks. Only a few have electricity or even running water. Families must pay for schooling and medical care. Unfortunately, the costs are too much for many, so they do without. In the northwest part of the country, there are some volcanoes. The sides of the volcanoes are covered in forest where about 250 gorillas live. Only a short time ago, there used to be over 300 gorillas, but they are being killed by poachers who want to steal their babies to sell to zoos. The forest is also threatened by overcutting of the trees. In just a few more years, the forest will be gone.

Outside of Rwanda and adjacent to its borders, there are still close to two million Rwandan refugees. The vast majority refuse to accept assurances from the international community and the Rwandan government that it is safe for them to return. They continue to fear being arrested as suspected genocidaries and thrown into one of Rwanda's notoriously appalling prisons, where they could languish for months until their case is brought to trial. Another reason many will not return, is that they enjoy a better quality of life in the refugee camps than do their counterparts living in Rwanda. However, the presence of these refugees, concentrated as they are into camps close to the country's border, is causing significant environmental damage.

The UN is not well liked by some Rwandans. It is important for us to remember though, that the UN was slow to intervene when so many people were being killed here last year. Rwandans are still distrustful of us, and only grudgingly accept our presence. However, despite the frustrations and the dangers, we must not allow ourselves to turn away from the challenges we face. I become concerned when I hear some Canadians complaining about our participating in UN missions because it is too expensive, too risky, or because they don't understand why we should care about problems in Rwanda or other places like Haiti and the former Yugoslavia. I can only respond that based on my personal experiences in Rwanda and other UN missions, we can't afford not to become involved. The Blue Beret with UN badge, and the distinctive red and white Canadian flag we wear on our shoulders, symbolizes Canada's long-standing and consistently demonstrative commitment to promoting peace and stability in the world's trouble spots and to safeguarding human rights. It is what we are known for the world over, and it is what helps to define us as Canadians. Peacekeeping is among the noblest of any nation's undertakings. I hope it is something that we will never allow ourselves to forget, nor cease to find reasons to keep doing.

I would like to think that we can stop the fighting here or anyplace where the UN has its troops, but conflict can only really end when the warring sides agree to resolve their problems, peacefully. Meanwhile, we try to deter the use of deadly force by any of the sides, and work to encourage a dialogue between them that will foster peace and stability. Sometimes it can be very exasperating work. The importance of any success is often measured against the efforts applied to achieve it. Consequently, in peacekeeping, even a little progress can be perceived as monumental and tantamount to having scaled Mount Everest to reach it.

We were able to follow the great referendum debate, but our circumstances here lead many of us to wonder why we can't accommodate each other in a country as rich in resources and stable as Canada is. We are a mixed group here: men, women, French and English speaking. We work with a unity of spirit and purpose. We have to. Because failure to do so would undermine the job that we have been sent here to do, and could jeopardize the safety and well being of each other. We recently had an occasion to sing Canada's national anthem as a group. It was spontaneous, and we sang it in French (those of us who didn't know all the words just hummed along). When we finished, there were not afterthoughts, save one. Whatever language we sang in, it was no less the national anthem of our country. I'm not sure what one can infer from this, but from my perspective in this reality, our Canadian-made problems at home quickly lose that precipitous aspect that we tend to award them, especially when one starts to make the inevitable comparisons.

Thank you for the card and letter. The card was certainly the most ambitious and striking of all those we received. I was compelled to carefully scrutinize the signatures to see if I could recognize any of the names. It is now prominently displayed in an area of our compound where it can be seen by all of us. Your support is heartening, especially at Christmas when our thought are of our families and friends at home. Merry Christmas and Happy New Year from Canada's contingent in Rwanda.

Somewhere in Rwanda
January 21, 1996

Dear Jane,

Your second letter arrived after only a week in transit. Not bad for such a long trip. Once again, I am reminded that many Canadian soldiers do indeed recognize their country's contribution to UN peacekeeping. I must add though, that UN service is not exclusively the domain of soldiers. As I'm sure you are aware, police and civilians drawn from all across Canada also participate in peacekeeping. In Rwanda, I have met Canadians working with the various aid agencies operating in the region, and also some who serve as UN human rights field officers. We form quite a community here in Rwanda.

Your peacekeeping "project" sounds intriguing to me. Having served on a number of UN missions, I am interested in knowing the thoughts of other peacekeepers. What I hear day-to-day from colleagues, I don't think really reflects their innermost thoughts. We, of course, react to the daily routine, the weather, sudden events, stress and traumatic experiences in a manner that is consistent with our natures. For some, that reaction is vocalized, for others it is not. Regardless of how that reaction is manifested, I think that it is later, when one addressed the issues retrospectively, that true feelings are expressed. I can imagine that many of your letters from peacekeepers portray thoughts, fears and feelings they have not shared with their comrades, or even their families.

The Canadian government, as you may be aware, has recently decided to terminate its participation in UNAMIR. Canadians can be proud of what our involvement has accomplished here. We did not solve all of the problems, but our presence has contributed to raising the quality of life for many Rwandans. I hope it was enough to sow the seeds of lasting peace. But, I am not naive. The issues that brought war, are not yet resolved. More work, more patience, more healing is needed. For the past few days in Kibuye, a small town along the shore of Lake Kivu, a mass grave has been undergoing examination by international experts. Around 200,000 people were believed killed in this area. In this one grave, there is thought to be up to 4,000 bodies. The skeletal remains provide more gruesome evidence of the horror that marked the genocide. Fighting in Burundi is now adding to the region's refugee problems. Some are now calling for a UN military presence in that country, but there is a reluctance to commit any more troops or money to this part of Africa. You might ask why Canada has chosen now to withdraw our contingent. I don't have the answer to that question. I just know that we did our best while we were here, and we made a difference.

I will take home many memories of my time in Rwanda, pleasant and not so pleasant. I will remember the unique beauty of the country. The people, who despite their ordeal, return with a smile with a smile, a wave with a wave. The friendships with members of other UN contingents. Among the more macabre: a battlefield strewn with the corpses of dead soldiers, a village where the lifeless bodies of men, women and young children executed in a revenge attack lay in their houses, the smell of death under a hot sun, the boobytraps and landmines that once again arbitrarily spare me, but take the man alongside. As I draw near to the end of a year on the peacekeeping road, one that has taken me to the Balkans and to Africa, I am eager to see home again. Meanwhile, many of my fellow soldiers from across Canada are on their way to other peacekeeping theatres. I wish them luck and a safe return.

I would like Canadians to recognize the importance of peacekeeping in today's world. It is a tradition worth maintaining despite the high costs we sometimes must bear. Once again, I appreciate your support.

I would like Canadians to recognize the importance of peacekeeping in today's world. It is a tradition worth maintaining despite the high costs we sometimes must bear. Once again, I appreciate your support.

◆ ◆ ◆

Haiti

The Canadian Contribution to Haiti

Mission Background

The presence of CF personnel in Haiti goes back to November 1990 when under Operation *Heritage,* 11 CF officers participated in the UN Observer Group for the Verification of the Elections in Haiti charged with verifying Haiti's election process. Canada also provided military advice on peacekeeping to the Organization of American States (OAS) in its efforts to return democracy to Haiti after the September 1991 *coup d'état* headed by Lieutenant-General Raoul Cédras that overthrew Haitian President Jean-Betrand Aristide. Early efforts by the UN and OAS to resolve the situation produced limited results. In June 1993, UN Security Council Resolution 841 imposed an oil and arms embargo against Haiti.

On July 3, 1993, President Aristide and General Cédras reached an agreement on the Haitian crisis. Under the Governors Island Agreement, President Aristide was to return to Haiti by October 30, 1993, and the leadership of the Haitian police and military were to retire. The agreement also called for the UN to assist in modernizing the armed forces and establishing a new police force in Haiti.

United Nations Mission in Haiti (UNMIH)

An attempt to establish the first UNMIH was made in September 23, 1993 under Security Council Resolution 867. Canada's initial contribution to the UNMIH was to involved 100 CF construction engineers to assist in rebuilding Haiti's public buildings and hospitals, and about 100 RCMP officers were to join the multinational UN civilian police force. Initially, a small number of CF and RCMP personnel deployed to Haiti as part of the original UNMIH advance party. After weeks of heightened violence and intimidation, armed gangs loyal to the regime of Lieutenant-General Raoul Cédras refused to allow the disembarkation of UNMIH personnel about the USS *Harlan County* at Port-au-Prince.

With the junta refusing to step down from office, it became clear that UNMIH could not deploy. The situation prompted the UN Security Council to approve

Resolution 875 on October 16, 1993, which called for the enforcement of sanctions against Haiti.

Initially, the Government of Canada authorized the deployment of a Canadian naval task group under Operation *Forward Action* (consisting of the destroyers HMCS *Fraser*, HMCS *Gatineau* and the replenishment ship HMCS *Preserver*) to Haitian waters. Alongside ships from the US, the UK, France, Argentina and the Netherlands, their mandate was to ensure cargo bound for Haiti did not violate UN sanctions. Several ships including HMCS *Kootenay*, *Annapolis*, *Ville de Québec* and *Terra Nova* served with the multinational Maritime Interdiction Operation.

UN Security Council Resolution 917 adopted on May 6, 1994 imposed far-ranging sanctions against Haiti. In July 1994, the Security Council authorized the formation of a multinational force to remove the Haitian military leaders and establish and maintain a secure and stable environment. In August 1994, Canada announced the deployment of 15 CF members to team up with counterparts from Argentina and the United States as part of a Multinational Observer Group (MOG). The MOG mandate was to monitor the border between the Dominican Republic and Haiti, and to provide technical advice to the Dominican Republic on the enforcement of the Haitian trade embargo.

On July 31, 1994, UN Security Council Resolution 940 authorized the formation of a Multinational Force (MNF) to "use all necessary means" to implement the Governors Island Agreement. It called for the deployment of 6,000 personnel as part of an expanded UNMIH, which would replace the MNF once a permissive environment had been established.

With the arrival of the American-led MNF on September 19, 1994, Haitian ports of entry and airports could be effectively monitored in a manner consistent with the relevant UN Security Council Resolutions. Consequently, HMCS *Terra Nova* and the MOG personnel in the Dominican Republic were withdrawn in October 1994.

In January 1995, UN Security Council Resolution 975 authorized the deployment of a 6,000 strong force. The transition from a multinational force to UN mission officially occurred in March 1995 and lasted until July 1996. Canada's contributed approximately 500 personnel along with 100 RCMP officers in support of the mandate of UNMIH which was to assist the Government of Haiti to: secure a proper environment for the fulfillment of the UNMIH mandate; professionalize the Haitian Armed Forces and create a separate police force; and conduct democratic elections.

President Aristide returned to power in October 1995. Under his guidance, free and fair elections (legislative and presidential) were conducted during late 1995. He disbanded the Haitian military and developed plans to improve the national police force and the Haitian economy.

The role of the Canadian Contingent in the overall UN mission was to provide helicopter support for the UNMIH force; a construction engineer company to work within the Force engineer battalion; and a medium lift helicopter platoon for the Force logistics battalion.

For the first time in Haitian history, the peaceful transition of power between two democratically elected Presidents, Aristide and René Préval, occurred on February 7, 1996. On February 9, 1996, the new Haitian President, Préval, requested that the UN extend the mandate for another six months.

UN Security Council Resolution 1048 adopted on February 29, 1996 extended and modified the UNMIH mandate until June 30, 1996. UNMIH was mandated to assist the Government of Haiti in sustaining the secure and stable environment established during the multinational phase and professionalizing the Haitian National Police. UN Security Council Resolution 1048 reduced the UN force to 1,200 military and 300 civilian police. Canada decided to provide an additional 700 military personnel in order to make UNMIH a viable military force for the tasks assigned under the new mandate.

United Nations Support Mission in Haiti (UNSMIH)

On May 31, 1996, the Government of Haiti requested the UN maintain a peacekeeping force in Haiti following the expiration of the UNMIH mandate. UN Security Council Resolution 1063 established on June 28, 1996, the United Nations Support Mission in Haiti (UNSMIH) with a mandate to assist the Government of Haiti in the professionalization of its police and in the maintenance of a secure and stable environment. UNSMIH also supports activities to promote institution-building, national reconciliation and economic rehabilitation. The initial mandate for the UNSMIH mission was for five months ending in November 1996. In December 1996, it was extended until May 1997. The UN funded 600 personnel of the military component of UNSMIH while the remaining 700 soldiers were funded by voluntary contributions from Canada (designated as Operation *Stable*) and the US.

The Force's peace support mission was to assist the democratic government of Haiti in fulfilling its responsibilities by: assisting in the continuing development and professionalization of the Haitian National Police force and related institutions; contributing to the sustainment of the relatively secure and stable environment that has now been established; and supporting activities to promote institution-building, national reconciliation and economic rehabilitation.

The concept of operations for UNSMIH's military component envisaged the continued stationing of a UN peacekeeping force in Port-au-Prince. This force maintained a visible presence throughout the remainder of the island by a regular patrolling programme. It was intended to assure the Haitian population of the UN's continuing interest and provide senior mission personnel with a means of assessing the effectiveness of the various programmes that were put in place to aid in the country's further development.

The Canadian Contingent in UNSMIH numbered 785 Canadians and included:

- a small reconnaissance battalion of about 350 personnel from the 5th Régiment d'Artillerie Légère du Canada (5th RALC) stationed at CFB Valcartier (a headquarters and two rifle/reconnaissance companies) manned and equipped to conduct patrols and other operations in Port-au-Prince and throughout Haiti;

- an engineer troop of approximately 70 personnel from the 5th Régiment de Génie de Combat (5th RGC) stationed at CFB Valcartier, including both field engineering and construction elements. This unit provided a range of capabilities such as route maintenance, vertical and horizontal construction, EOD, water supply, power generation and other engineering services;

- a utility tactical transport helicopter squadron of about 85 personnel mainly from the 427th Squadron stationed at CFB Petawawa equipped with five CH-135 Twin-Hueys (a sixth aircraft was held as reserve) and two Russian-built Mi-8s (under UN contract) provided the UN force with: casualty evacuation, a day/night mission capability and a medium airlift transportation capability;

- a military police platoon of 16 CF personnel including the platoon commander/ Force provost marshal provided a criminal investigation capability to the Force Commander;

- a Military Information Support Team (MIST) of 24 CF personnel provided timely and accurate passage of UN information to the civilian population;

- a Canadian Contingent Logistic Group of approximately 190 personnel was responsible for the provision of virtually all support services required to administer and sustain Canadian units deployed with UNSMIH. It included a maintenance platoon, a logistics platoon, a transport platoon, a medical platoon and an administrative and headquarters platoon. It is also providing supervisory capability to the UN heavy workshop and transportation sections; and

- Canada provided 50 CF personnel to serve in various capacities within UNSMIH headquarters.

United Nations Transition Mission in Haiti (UNTMIH)

On July 29, 1997, the UN Security Council adopted Resolution 1123, which established UNTMIH. The mandate of UNTMIH was to assist the Haitian government in its continuing efforts to improve the Haitian National Police. The role of the military component of UNTMIH was to ensure the safe and free movement of UN personnel carrying out the mandate. UNTMIH was limited to a single four-month period which ended on November 30, 1997.

The Canadian Contingent in UNTMIH numbered 650 Canadians and included:

- elements of the 2nd Battalion, R22eR from Quebec City which conducted patrols and other operations in Port-au-Prince and throughout Haiti;

- a helicopter squadron from the 430th Tactical Helicopter Squadron at CFB Valcartier, which provided the UN force with casualty evacuation, a day/night mission capability and a medium airlift transportation capability;

- a military police platoon from CFB Valcartier, which provided criminal investigation services;

- a MIST from the 5th Canadian Mechanized Brigade Group (5th CMBG) stationed at CFB Valcartier; which provided accurate UN information to the population;

- a logistics group, from the 5th CMBG, with elements from air and maritime units across Canada, which provided support services to administer and sustain Canadian units deployed with UNTMIH. It included a maintenance platoon, a transportation platoon, a medical platoon, an administrative and headquarters section and an engineer troop; and

- more than 50 Canadian civilian police officers participated in the mission as part of the UN Civil Police contingent.

Humanitarian Projects

During the operation, Canadian military personnel:

- built four schools in the hills of Haiti;

- made more than 115 school desks and chairs;

- painted, renovated and repaired many roofs, walls, windows and doors in orphanages and schools;

- installed sinks and plumbing for a well;

- laid out and fitted several soccer fields and other play areas;

- repaired a Red Cross ambulance and several other vehicles;

- operated several mobile medical clinics;

- distributed more than 12,000 pounds of non-perishable food and 342 litres of infant formula;

- donated more than 2,000 toothbrushes;

- distributed 15 bales (6,000 pounds) of clothing plus 115 other crates of clothes and shoes; and

- distributed 50 boxes of school supplies.

Key Events involving the CF in Haiti:

- In February 1996, the Canadian government announced Canada would take command of the UN forces in Haiti, under the command of Brigadier-General Pierre Daigle.

- From July 1996 to July 1997, Canada led the UNSMIH

- From July 1997 to November 1997, Canada led the UNTMIH.

- On August 29, 1997, Canada handed over the task of guarding the National Presidential Palace to the Haitian National Police.

- In early September 1997, Canadian divers helped recover many of the victims of the overturned ferry *Fierté Gonavienne*, until relieved by US Coast Guard divers. The Canadian Contingent provided logistical support to the American divers during their operations.

- On November 30, 1997 the mission officially stood down and ceased operations, and the personnel returned to Canada.

- From December 1997 to the present, Canada has seconded 12 civilian police officers to serve in the newly established UN Civilian Police Mission in Haiti.

This section is based on public information documents produced by the Canadian Department of National Defence Office of Public Affairs.

Letters from Haiti

The following letters detail the correspondence of three men and women (of varying ranks and occupations) who served in various aspects of the Haitian mission previously mentioned and who wrote to Jane Snailham from April 1995 to September 1995.

The following letters are three of seven written by a 44-year-old chief warrant officer who served at the headquarters of the Canadian Contingent assigned to UNMIH:

Camp Canangus, Port-au-Prince, Haiti
April 18, 1995

Dear Jane Snailham,

I am the Canadian Contingent chief warrant officer (CWO) of Operation *Pivot* in Haiti. I work directly for the Contingent Commander and I am responsible to him for the overall discipline, conduct, control, morale and welfare of the non-commissioned members in Haiti. We total 474 Canadians of diversified trades including logistics, pilots ([flying] twin Huey's) ground crew, engineers, administrative, finance clerks, postal, press, public affairs and all these personnel are from units across Canada. It is the first time Air Command has been tasked to provide a full Canadian Contingent for UN operations. We are quite proud of that fact!

I am 44 years old and have been in the Canadian Forces for 24 years. I'm married and my husband is in Petawawa. He's retired and attending school as the present time. I have no children, but I do have a spoiled gray tabby cat. My parents live in Petawawa, therefore, my husband visits them quite often.

I arrived in Haiti [in] March 1995, and my tour should end in October of 1995. Haiti is extremely hot and the Haitians are very poor. They're very proud people and quick to smile if you greet them in a friendly fashion. Their language is Creole which is very similar to French and since I am bilingual I'm able to converse with them quite easily.

We are situated at Camp Canangus which is close to Port-au-Prince. I'm certain [that] if you look at a map of Haiti, you'll notice an airport near Port-au-Prince, and we're just off the ramp. We were in tents for the first three weeks but now most of our troops are in makeshift buildings. We now have Canadian cooks, therefore, the food has improved, but the water is contaminated, therefore, we drink bottled water. We have portable toilets and showering facilities. The living conditions have improved since we moved in so the troops are relatively content. We assumed control from the MNF on March 31, 1995 and declared operational readiness. This past month has been a very busy one for us, but we're working together and all is going fine.

We had some difficulty with mail going in and out of the country, but I think it's sorted out now, so you should receive this letter promptly. Say hello to your husband for me, and thank you again for the lovely letter. It's nice to know there are Canadians who care and support us. It's difficult for Canadians to understand sometimes that military personnel are human beings with weaknesses and strengths that are trying to do a good job in diversified roles.

I'm honoured to write you letters and invite you to experience the next six months in Haiti. Take care of yourselves and I'll keep in touch. All the best until next time.

——————————— ◆ ———————————

Camp Canangus,
Port-au-Prince, Haiti
May 12, 1995

Jane,

Hello pen friend! I loved reading your letter — I couldn't believe Canada still had snow in April. I thought April showers bring May flowers. Ha! Ha!

We're fine and doing well. Some of us have started teaching classes for the Haitians. We teach them English and French and they're teaching us Creole. There are approximately 30 classes ranging from basic level to advanced. They love to learn and love hearing about Canada. They don't believe we have snow and that it gets cold in the winter. I have a class with 10 students and they're pretty happy and always smiling.

Many Canadians are working on projects for the missions in conjunction with our taskings and enjoy interacting with most of the Haitians. It's easier for Canadians as many of us speak French.

The weather here has been great, except for the evenings when it rains. Unfortunately, the ground gets pretty muddy and it isn't too great for the boots. Lately, we've been heavily involved with support missions for other UN troops in a transport, supply and engineering capacity. I had an executive meeting with my counterparts in the Nepalese, Honduran, Guatemalan, Pakistani, Bangladeshi,

American, Indian and Caribbean Community battalions in Haiti, and it was interesting to share information and perceptions. It's so interesting to meet troops from other countries with different traditions. It's an experience I could never again recapture in my lifetime. The colonel and I have been invited to their camps, so we definitely plan on going.

We also have started our sports programs. We have weights, stationary bikes, a volleyball court and we're working on soccer and baseball [fields]. We still have to level some of the ground and get gravel, but, hopefully, that goal will be attainable in the next couple of months.   My hobbies and interests are camping, canoeing, hiking, jogging, light weight training, aerobics, speed-walking and reading. I love to write as well. My husband always says he can't keep up with my letters.

I've taken over the job of doing human interest articles for various newspapers in Canada, on the troops in Haiti. The pictures and articles will appear in diversified base newspapers, and the troops seem pleased when they're chosen for an interview. I haven't had a person from the east coast yet, most of them were [from central or Western Canada such as] Moose Jaw, Comox, Edmonton, Cold Lake and Winnipeg.

The colonel said to say hi! Take care of yourself Jane. Bye!

Camp Canangus,
Port-au-Prince, Haiti
May 17, 1995

Jane,

I've enclosed some photos of our makeshift church, recreation centre, tent lines and aircraft and hangar lines. The camp is quite large and we've been busy cleaning up and building surrounding areas to make life more comfortable for the Canadians and Americans.

Today, I made a short video wishing Canada a happy Canada Day from UNMIH. Hopefully, you'll see it on Canada Day, as my troops were so happy to be able to say hi to their families at home.

The next three weeks are very busy for us, in conjunction with our taskings, we have many very important visitors such as the commander of Air Command and the chief of defence. We're very excited and pleased that they'll be visiting us in Haiti and all the units will be debriefing the VIPs on their roles and task assignments. I'll try to send some pictures of the military dignitaries with the troops, as I'm certain you would appreciate that.

I'm fine and doing very well. We're very busy but happy and are enjoying the mission. The days just fly by and it's great just mingling with the other soldiers like the Nepalese, Bangladesh, Americans, etc. It's quite an experience as every culture has its different habits and preferences from religion to food.

The colonel is doing just fine, but is very busy. Our taskings have increased extensively so we're kept hopping, but the troops always rise to the occasion. Busy troops are happy soldiers. Many times, I sit back and wonder how everything is going in Canada — I definitely can't believe you've had snow in May. Wonders never cease! We're making up for it here, as it's hot, humid and sunny. It rains at night sometimes, but not very often.

I will be proceeding home for a visit and I'm definitely looking forward to seeing my husband for a couple of weeks. My parents are back from Florida now, so I may see them when I visit Canada. I know it's a couple of month away, but I'm still excited.

Well, I'll sign off for now, Thank you for your lovely letters of encouragement. The colonel and I read every one. I will keep the picture as I've talked to my husband about my pen friend and I know he'd love the picture. Take care and I'll write again soon.

◆　◆　◆

The following letter is the first of three written by a corporal who served with a communications unit assigned to the Canadian Contingent of UNMIH:

Camp Canangus,
 Port-au-Prince, Haiti
May 23, 1995

Dear Jane,

I received your letter about 10 days ago, but I didn't even see the time go by — I guess they do keep us busy. Once we got in our little routine, a day looks like the other one before, and the next thing you know a week has gone by — it's okay though cause I don't mind time going by fast.

Now the next step to look forward to are my vacations. They're coming up already! I was supposed to have my next of kin flown down here to meet me. But that won't happen.   They have this conflicting policy and the fact that I'm single, got me at a disadvantage.   I guess my sister will be disappointed to lose her free airfare, but instead, the military will fly me back to Canada. I really don't mind going back. I'm actually probably going to enjoy this more. So mid-June, I'll get to see friends and family again for two weeks. I'll get to taste a bit of freedom and civilization! I'm getting more and more excited just thinking about it!

This is what I do — work with satellites. I'm part of the communications. I belong here to National Rear Link. We are the link back to Canada. A bit under the modem principle, we send messages back and forth. We are also in charge of the telephone system down here in Haiti (within the camp) and back to Canada. For instance, if I were to pick up the phone here to call Canada, the signal would go to a big satellite dish out here; it would shoot to a satellite, then back down to [a relay station in] Quebec. From there, the signal goes through special lines directly to Ottawa. From Ottawa, the phone call is forwarded wherever via normal Bell Canada lines. All that distance gives a bit of a delay when we talk on the phone.

So really, I can say I do a desk job. Once everything is set up, [which takes]

about two weeks of hard work, we show up for our shift. We work 24/7 [24 hours a day, seven days a week]. So two days, two nights, two evenings and then I get two days off to try and catch up on some sleep. I find it rather difficult to switch like that from day to night, but that way no one gets stuck with the same shift for a long time. We get the messages and file them and distribute them. Nobody even notices we do our job until the phone system goes down or until a message gets lost.

We are sort of confined to camp (unless we have some official business down town or unless we are on our days off) — in that case they drive us to a specific beach and we have to stay there. I think they want to minimize the contacts with the outside world for our own safety. It's not a danger zone like Yugo or Rwanda, that's for sure, but people here are starving. They know we have money and they know the UN has food. There is also a bit of an anti-UN movement going on — nothing big. But people here are a bit disappointed with the time it takes to rebuild a country. They were all hoping for jobs I guess. It is just not going to happen this fast.

I would say that most of the people are thankful and respectful, but, there are always thugs and desperate people, like anywhere else. And of course, you have some rich people that are not that happy to see the UN here. They had some power before and they would obviously prefer keeping it, but democracy sometimes goes against their plans. These people are going to try to discredit the UN and our work. That's something we have to deal with, but, like I said, from my own personal experience, people are nice and very friendly.

Once again, we don't have much contact with the outside world, I actually do my best to get implicated in different activities. I have a rotating schedule which sometimes gives me a day off. So I have to go on a few missions with the "padre," our priest, to orphanages and stuff like that. That gives me a direct opportunity to meet real people and to have a real contact with Haiti. The other Haitians we meet are all people hired by the UN, so of course they have been screened and of course they don't necessarily represent the real world.

I also have this great opportunity to feel useful and to feel like I'm doing more than handling messages. I'm teaching English to Haitian people, an hour per day. It is great. [The school I teach at] is called the School of Hope, and there are about 300 students. I have 10 in my class and they range from 19 to 40 years of age and from total illiteracy to a fairly fluent English. I have had a lot of teaching experience before so all that experience is handy. Most of them speak French, so it easy for me to explain stuff to them. But for the ones that only speak Creole, I get some help from other students, I'm actually picking up Creole pretty fast.

A very rewarding experience, but also very difficult and demanding. It is a difficult cause, I am put face to face with poverty and misery on a daily base. I have students that are literally starving, One of them can barely sit since she is so weak. It is sometimes a bit depressing to see how difficult their life is. They try very hard and they are eager to learn.

I find it funny cause before I joined the military, I knew virtually no English at all. I knew basic English. When I went to my military training (what some call boot camp), I was plunged into a totally English surrounding. I wanted to learn English and I figured it would be the best way to do so. I just had no choice, but to learn! I made lots of friends that summer and they made lots of fun of me and my English. It took me a week to start understanding people. But I learned! And now, if these friends of mine, knew I was teaching English! Boy, did I ever "move a lot on the monopoly board of life" in the last five years.

On a totally different topic, my co-workers here also planted a garden. They are going to see how well things grow in Haiti. It might actually be too hot for some of the things they planted here. But we'll see how it goes.

Maybe next time I'll write to you about our beach trips. I guess we do get an incentive to coming down here. They bring us to a place called Wahoo Bay beach and it's just great to have this opportunity to get away from it all and to relax a bit.

Thank you very much for your letter. It was enjoyable to hear from other thing than our day to day worries here.

◆   ◆   ◆

The following letters are the remaining four of the seven written by the 44-year-old chief warrant officer who was mentioned earlier on pages 159-162, and who served at the headquarters of the Canadian Contingent assigned to UNMIH:

Camp Canangus,
Port-au-Prince, Haiti
June 16, 1995

Jane,

I'm fine and doing well. We're keeping busy as the Phase One elections are in progress. The ballots are here in camp and soon our aircraft and trucks will be disseminating the votes.

Not all the very important persons visited as the Bosnia situation necessitated the cancellation of the chief of defence staff visit. It's unfortunate, but we understood as our cohorts in Bosnia are having a very difficult time. Thus far, our engineers are setting up camps for other UN battalions and they're doing well.

Many camps here have been completed, but the engineers have some more to build. After they've completed the camps, the engineers will commence humanitarian work. The engineers will enjoy that as my visit in Les Cayes [a coastal city on southwest of Port-au-Prince] revealed they had fixed a bridge and a school for the Haitians. The Haitians were definitely pleased to see the commander and his entourage.

I definitely enjoy the new experiences here in Haiti. Unbelievable as I'll never experience the situations I am involved in again in my career. We're lucky in the military as we visit other countries and get to meet all types of people from different countries.

I will be going home for two weeks leave at the end of July and I'm looking forward to the break as I'm tired.

I will be posted to a new position on my return to Ottawa, so I'll have more change in my life. I believe my husband will move to Ottawa with me, as it's difficult to maintain a relationship when husband and wife are in two locations. This transfer will mean I will be in Ottawa for another three years. Luckily, I like Ottawa, so time should fly by.

Well Jane, I've gabbed enough, so I'll close for now. Take care of yourself and keep a stiff upper lip. Bye.

◆

Camp Canangus, Port-au-Prince, Haiti
July 3, 1995

Jane,

Hi there! Sorry I haven't written lately, but we've been busy with the elections in Haiti and visitors from all over. This week is another busy week as the Americans will be celebrating the Fourth of July — of course we had a show tour from Canada for the First of July. We invited all the commanders from the different countries, the Canadian ambassador and our Force Commander. The Canadian flags were flying and I was very proud I must admit.

Haiti is the poorest country, therefore, the Haitians appreciate any and all assistance provided. The criminal activity is high, but we're not in a war zone like in Yugoslavia. The Haitians are just concerned about their next meal. Hopefully, you'll be able to bring it together.

The students I was teaching have graduated and now have a Haitian teacher. Now the contingent is liaising with the missions and providing scrap wood for shelters some used clothing and any extras that are available. The engineers have done some small projects in the villages, but funds and materials aren't available for larger projects. We're not certain if funds will be allocated for humanitarian jobs as the elections are in full swing and all UN peacekeepers are busy securing and delivering voting ballots so fair elections can proceed. It's not easy as Haiti is a large island with minimal roads if you can all them roads. Fortunately, the helicopters are working overtime to deliver support equipment, ballots, UN troops, etc.

We're all fine here and looking forward to a short break after four months of full tilt. On July 24th, I go home for 14 days on UN leave and I'm glad as I'm pretty tired. The colonel goes on leave on July 9th and he definitely needs to relax, as the pace lately has been extreme. The heat is still pretty high here, but most people are acclimatized, except for the Americans who just rotated with new troops. We have a new American officer in charge of the American helicopters — he seems pretty good so that helps. They're adjusting to the Canadian way of doing business, but it is difficult to orientate them to the Canadian way.

Well, I've run out of things to say and I have other letters to write, so I'll sign off for now. Take care of yourself. Bye!

------------------------- ◆ -------------------------

Camp Canangus, Port-au-Prince, Haiti
July 7, 1995

Jane,

Hi! I enjoyed reading the article [on you] from the Defence Report — Department of National Defence Public Affairs Office, Atlantic. Quite impressive and obvious that you are a very rare lady who is actively involved with what the military peacekeepers are doing. Our roles are diversified based on the area we are serving in, but you appreciate the difficulty and personal hardship most troops encounter.

We were heavily involved with the elections as all the ballots were brought to Camp Canangus. Our vehicles, helicopters and American[-made] [Hercules transport aircraft] delivered the ballots to diversified locations in Haiti. The Americans and Canadians worked closely together to ensure all went well. The RCMP were involved in the security of the voting sites and were kept very busy. We still have Phase Two to go through in July/August and then the presidential election is the final step.

Right now, the engineers are heavily involved with more camp site set up for the UN infantry battalions within Haiti. We're also busy doing some humanitarian projects for a few villages and the missions. We're not able to do major projects until the mandate changes, as the humanitarian assistance is not the main mandate.

We've been very busy lately as we're in the process of preparing a second contingent for rotation in September. By then it'll be six months that my group is in-theatre. The new commander and his CWO came for a week to get an idea of what Haiti is like. They were pretty impressed with the camp, as all they have to do is walk in and take over. Most of the work is done and all they have to do is maintain the status quo. The first group in, normally has the most work.

I've been posted to a new position in Ottawa, effective October 1, 1995. My husband is presently in Petawawa, so I'm not sure if he'll move with me. I may have to continue to commute for another year.

The commander is fine and proceeds to Colorado Springs, [Colorado, USA] on leave, on Sunday. He'll be meeting his wife there, which is good — I must admit we're tired. On his return, I will be going home to Canada for two weeks. By then, I'll need a rest.

Well Jane, I've run out of things to say, so I'll sign off for now. Take care of yourself — say hi to your family for me and thanks for your continued support.

Camp Canangus, Port-au-Prince, Haiti
July 18, 1995

Jane,

Hi there! I received your letter and was pleased to hear about your invitation to the International Tattoo. I can't think of anyone who is more deserving than you are. By now, the evening is over, but life is full of fond memories and I know you appreciated the evening.

The troops in the military are very proud to wear the uniform of the CF and regardless of the negative press most of the soldiers do their best regardless of the difficulty of the mission. You're one of the rare Canadians who have insight into the military way of life.

I just finished a visit with the Chief of Defence Staff General de Chastelain and the troops were very happy he came to see us. General de Chastelain has had his share of bad press and it's unfair as he cares about the basic soldier and listens to us when we talk. It was a busy three days, but the general visited all the troops and thanked us all for our hard work. It meant a great deal to us that he took the time to visit as he has a hectic schedule.

I received the UN medal for the mission in Haiti and on July 21st I will be conducting a parade whereby all the troops will be presented their medals. It'll be a special day for us and there will be many proud faces on parade. There are compensations for a job well done!

I'll be flying to Ottawa on July 24th, and my husband will be meeting me at the airport. We'll be staying in Ottawa for a few days and then we proceed to Petawawa. I plan on getting some rest for a few days, and my husband and I have some work to do outside the house. I return to Haiti on August 7th, and the last month-and-a-half will definitely be very busy.

Well, I'll sign off for now. Take care of yourself Jane, and I'll keep in touch. Say hello to your family for me, and I'll tell the colonel your prayers are for him as well. Bye for now.

◆   ◆   ◆

The following letter is the second of three written by the corporal who was mentioned earlier on pages 162-164, and who served with a communications unit assigned to the Canadian Contingent of UNMIH:

Camp Canangus, Port-au-Prince, Haiti
August 21, 1995

Dear Jane,

Well, I think this is about it!

Officially, today, I start my count down towards the end. In 10 days, I'll be leaving this country. It's hard to describe how I feel about it.

I think I can look back and say that I had a good tour. I had a good tour for the country and for Canada, of course. I did my job and more. I gave all that time to the School of Hope and I'm pretty sure it made a difference. So, I did something good for Haiti. But on top of that, I didn't break anything expensive, so for Canada it was also a great tour. But, more important, for me too. I was fearing a bit a UN tour. A tour is such a long time, so I was fearing this a bit. And it turned out good. No major depression, no big time blues from home. I didn't hate it. I actually enjoy most of it.

I got lucky. I was here with good people. I had good tent-mates and a good tent that kept bugs and dust outside. I had a good camp with good facilities and easy access to good food. Over all, this was an awesome tour. I managed to keep myself busy most of the time.

Finally, after over a year-and-a-half, I can see the end of the tunnel. I am about to get control over my life again. That was what I missed the most. I wanted to be the one controlling my life and deciding for myself, not the military anymore. So I should be really happy. Finally, able to see friends and family again. Finally, being able to drive around and live in the "best" country in the whole world! Finally, back to school. Isn't this all I've been dreaming about for so l

So I guess I should be extremely happy to go home. Frankly, after all, I was getting a bit tired of this place. It was a good tour, but it came with it's truck load of frustrations as well. So I guess this should be celebration time! But it's not quite. It is weird to leave this place behind. I look around and try to see a positive visible change in the country and I don't. We must have achieved something. I know we have! But can I see it? What has changed since the beginning of this adventure? I know I have. I am I think, a bit better of a person. I know a bit more about myself and I appreciate Canada a whole lot more. I am this much more ready to face life, but that's not why they sent me here. Did anything else change? Can I do something else before I leave within ten days.

I know that my students (the initial ones from the month of May) are going to graduate this Saturday. That is something I should be proud of (so should they). I don't know if it'll help them in their day to day life, but we did reach our goal together. What else can I do in the next 10 days to do that little extra thing.

How about all these military people that I've lived with for the last little while. It is kind of weird to each go our way. Specially, the one I actually got to know. I have these new American friends that I don't really want to leave behind (they still have 55 days left here).

It's weird — to be hesitating about going back. After all, for the last six months that is what kept me going — the fact I was going back to Canada eventually. And still now, it's hard to face the fact that it's all over. As soon as I step in the plane, as soon as they hand me those peanuts and as soon as I put my headset on and start watching the movie, everything from here will stay behind and vanish towards the state of memories. Will all this still keep going on, or was that all a big illusion?

I actually fear a bit the return to reality. Nobody will really know what it was and how it is here. No matter how long I talk about it. And soon enough, I'll have to start preoccupying myself with other stuff. I'm not exactly looking forward to the routine that is waiting for me. I think I had started to build too big expectations about what it was going to be back in Canada. I'm not sure my student life can compete with the illusions I had built about my life in Haiti.

Then again, I think for a little while, life will be so sweet in Canada. The simplest things over there, that people take for granted, will be so sweet. Rediscovering life, Ottawa, Canada, Canadians (the best people on the planet). I now know what I missed about Canada and what made Canadians so great to me. Therefore, I now know how to make Canada an even better place and how to be a better Canadian myself. I know how I don't want things to be.

Thank you for your support and prayers. Thank you for your mail and attention from home. It is very much appreciated. I feel very lucky that I'm the one who picked up your letter here in Haiti. I thank you again and hope that you find time to keep writing to other peacekeepers like you have been doing for so long. People appreciate that a lot. It is your contribution to peacekeeping missions and you should not underestimate it. I wish you and your family the best.

◆　◆　◆

The following letter was written by a soldier who served with the finance section of the Canadian Contingent assigned to UNMIH:

Camp Canangus, Port-au-Prince, Haiti
September 24, 1995

Hello Jane,

It was so nice to receive your letter the other day. Of course, all mail from "home" means a lot, especially one as supportive as yours.

I'll tell you a little about myself. I've been married for the past 13 years. My husband is also in the CF and we met when we were both stationed in Bermuda. We were recently transferred to [CFB] Trenton, Ontario where he has set up home. We have two boys, aged nine and 10. They are quite a handful as two boys can be. But they both do very well for having a set of military parents. I've been in the CF for 20 years now, I joined right out of high school at the age of 17. I have been stationed all across Canada and as far south as Bermuda. I have also been to Africa (in 1993) working with the air force flying food relief into Somalia from Kenya. My primary job is as a finance clerk. This encompasses paying the troops, paying the bills and keeping track of the money. A popular job with the troops. I usually try to get involved with other things, though, especially when I'm on a tour like this.

I love being part of the military and I especially like doing this type of thing. Although I miss my family terribly, I joined the military to hopefully do some good in the world and these missions allow me to perhaps do that. I feel it is also somewhat part of my upbringing, as my father was in the British [Royal] Air Force for 25 years, my mother for five years, my sister is currently serving and also posted in Trenton, and my brother spent several years as an Air Cadet.

Haiti is quite a different experience. The poverty is overwhelming, but I find they do seem to have more concern for human life than the people in Africa did. Unfortunately, on many occasions these people do not have the knowledge or equipment to help themselves. With time perhaps, we can help them with that.

Well, must go now as I want to drop a line to my family tonight. I must thank you again for all your support. I know I speak for many peacekeepers when I say I wish that there was more people like you, who show their support for our endeavours. Take care of yourself.

◆　◆　◆

The following letter is the last of three written by the corporal who was mentioned earlier on pages 162-164 and 168-169, who served with a communications unit assigned to the Canadian Contingent of UNMIH. This letter was written after the soldier rotated back to Canada:

Postscript, Gatineau, Canada
October 10, 1995

Dear Jane,

I certainly should have written to you a long time ago. I have received two letters from you here in Gatineau. I guess I got sort of caught up in the routine and school work. I think I can say now that I am mostly adapted to the real life. It was a bit more difficult than I thought it would be.

I had lots of dreams about myself being back in Haiti. I wasn't sad about it, or mad or anything. In every dream, I kept coming back to the same conclusion, that this was the place for me to be and that I had still so much stuff to do over there. So within even dream, I ended up getting back easily in the routine again. There was always so much to do.

I often think about Haiti, and I sometimes that it is actually the place where I should be. It is weird, I am so happy to be back, but at the same time, I have the impression that there is lots left for me to do there. I know I have done my share and I should let other people do stuff too, but it almost felt [like] home over there. I felt like I had good reasons. I have to do stuff for myself. I go to school for myself. It is weird. It doesn't seem to be a good reason enough to do stuff. It feels like I belong over there. And sometimes I feel bad about the fact that I am not there any longer.

I try my best to use logic to get over these thoughts. After all, I did more that what I had to do and now it is time to think a bit about myself, I think.

Take good care of yourself and your family.

The Edge of Sanity

In the building high, an office bright,
I see the ravages of war —
A peacekeeper's slides tell the story
Of this horrific UN tour.

My mind is troubled, my heart cries,
The images entrenched in my mind —
A war torn land, a people grieve
their lives changed forever.

The slide clicks over, a face appears
Lined, wrinkled and old —
She wipes a tear from her eye,
The camera sees it all.

War strips her of her heart, her life,
her dignity, her self,
Pain is all she feels —
I want to hug her, take her away
To safety, to peace.

Bullet holes, destroyed homes
Buildings barely there —
A mortar here, a sniper there,
Survival, a game of chance.

Several children sit on a wall
Like any child will do —
But here?

(DND photo)

How can they smile?

How can they wave?

At a lens focused on them.

My heart cries, I send my prayers

To this tortured land,

That one day soon the warriors will say,

Let's give peace a chance.

Jane Snailham
March 4, 1994

(DND photo)

Glossary of Terms

2nd CER	2nd Canadian Engineer Regiment
3rd CSG	3rd Canadian Support Group
4th CER	4th Canadian Engineer Regiment
5th RGC	5th Regiment de Génie du Canada
8th ACCS	8th Air Communication and Control Squadron
12th RBC	12th Regiment blindé du Canada
95th CMSG	95th Composite Mission Support Group
95th FLSG	95th Force Logistic Support Group
ALCE	Aircraft Logistic Control Element
AOR	Area of Responsibility
ARRC	Allied Command Europe Rapid Reaction Corps
BBC	British Broadcasting Corporation
BC	British Columbia
CANBATT One	Canadian Battalion One (Croatia)
CANBATT Two	Canadian Battalion Two (Bosnia-Herzegovina)
CANLOGBATT	Canadian Logistical Battalion
CBC	Canadian Broadcasting Corporation
CCSG	Canadian Contingent Support Group
CDN	Canadian
CF	Canadian Forces
CFB	Canadian Forces Base
CMAC	Cambodian Mine Action Centre
CMBG	Canadian Mechanized Brigade Group
CMED	Central Medical Equipment Depot
CNN	Cable News Network
CO	Commanding Officer
CPAF	Cambodian People's Armed Forces
DCO	Deputy Commanding Officer
DPA	Dayton Peace Accords
ECMM	European Community Monitor Mission
EOD	Explosive Ordnance Disposal
FAC	Forward Air Controller
FMED	Forward Medical Equipment Depot
FYROM	Former Yugoslav Republic of Macedonia
HMCS	Her Majesty's Canadian Ship
HQ	Headquarters

IDP	Internally Displaced Person
IEBL	Inter-Entity Boundary Line
IFOR	Implementation Force
IPTF	International Police Task Force
JNA	Yugoslav Army (Serbia and Montenegro)
KR	Khmer Rouge
LdSH(RC))	Lord Strathcona's Horse (Royal Canadian) Regiment
MASH	Mobile Army Surgical Hospital
MIST	Military Information Support Team
MND	Multinational Division
MNF	Multinational Force
MOG	Multinational Observer Group
MP	Military Police
MSF	Medicins Sans Frontieres — Doctors Without Borders
NADK	National Army of Democratic Kampuchea
NATO	North Atlantic Treaty Organization
NB	New Brunswick
NBC	National Broadcasting Corporation
NCO	Non-Commissioned Officer
NGO	Non-Governmental Organization
NS	Nova Scotia
NSE	National Support Element
OAS	Organization of American States
OIC	Officer in Charge
OP	Observation Post
PAO	Public Affairs Officer
PEI	Prince Edward Island
PGM	Precision-guided munitions
PMQ	Private Married Quarters
PO	Petty Officer
PPCLI	Princess Patricia's Canadian Light Infantry
R22eR	Royal 22nd Regiment
R&R	Rest and Relaxation
RCI	Radio Canada International
RCD	Royal Canadian Dragoons
RCMP	Royal Canadian Mounted Police
RCR	Royal Canadian Regiment

ROWPU	Reverse Osmosis Water Purification Units
RPA	Rwandan Patriotic Army
SOC	State of Cambodia
SFOR	Stabilization Force
STANAVFORLANT	Standing Naval Force Atlantic
UK	United Kingdom
UN	United Nations
UNAMIC	United Nations Advance Mission in Cambodia
UNAMIR	United Nations Assistance Mission for Rwanda
UNICEF	United Nations Children's Fund
UNHCR	United Nations High Commissioner for Refugees
UNMIH	United Nations Mission in Haiti
UNMO	United Nations Military Observer
UNOMUR	United Nations Observer Mission Uganda-Rwanda
UNPA	United Nations Protection Area
UNPF	United Nations Peace Forces
UNPROFOR	United Nations Protection Force
UNREO	United Nations Rwanda Emergency Office
UNSMIH	United Nations Support Mission in Haiti
UNTAC	United Nations Transition Authority in Cambodia
USA	United States of America
ZAMBATT	Zambia Battalion
ZOS	Zone of Separation

THE CANADIAN PEACEKEEPING PRESS

The Canadian Peacekeeping Press, the publishing arm of the Pearson Peacekeeping Centre, has a number of publications of interest. These include: *UN Peace Operations and the Role of Japan* (edited by Alex Morrison and James Kiras, softcover, 123 pp., $20 + taxes and shipping); *Facing the Future: Proceedings of the 1996 Canada-Japan Conference on Modern Peacekeeping* (edited by Alex Morrison, Ken Eyre and Roger Chiasson, softcover, 222 pp., $20 + GST and shipping); and *Seeds of Freedom: Personal Reflections on the Dawning of Democracy* (by Senator Al Graham, softcover, 288 pp., $30 + taxes and shipping).

Recent publications of the Canadian Peacekeeping Press include: *Theory, Doctrine and Practice of Conflict De-escalation in Peacekeeping Operations* (by David Last, softcover, 152 pp., $23.50 + taxes and shipping); *Refugees, Resources and Resoluteness* (edited by Alex Morrison, Stephanie A. Blair and Dale Anderson, softcover, 186 pp., $20 + GST and shipping); and *Multilateralism and Regional Security* (edited by Michel Fortmann, S. Neil MacFarlane and Stéphane Roussel, softcover, 262 pp., $23.50 + taxes and shipping).

To obtain a Publications Catalogue or for more information on the Canadian Peacekeeping Press, please contact:

Sue Armstrong
Publications Manager
Pearson Peacekeeping Centre
Cornwallis Park, PO Box 100
Clementsport, NS B0S 1E0
Tel: (902) 638-8611 x 161
Fax: (902) 638-8576
Email: sarmstro@ppc.cdnpeacekeeping.ns.ca
Website: http://www.cdnpeacekeeping.ns.ca

THE CANADIAN INSTITUTE OF STRATEGIC STUDIES

Chairman of the Board of Directors, Hon. Jean Jacques Blais
President, Alex Morrison, MSC, CD, MA
Executive Director, David Rudd, BA, MA

The Canadian Institute of Strategic Studies (CISS) provides the forum and is the vehicle to stimulate the research, study, analysis and discussion of the strategic implications of major national and international issues, events and trends as they affect Canada and Canadians.

The CISS is currently working independently or in conjunction with related organizations in a number of fields, including Canadian security and sovereignty; arms control and disarmament; Canada-US security cooperation; Maritime and Arctic security; Asia-Pacific security studies; social issues such as drugs, poverty and justice; national and international environmental issues; and regional and global trade issues.

CISS publications include:

Free with membership:

The Canadian Strategic Forecast
Seminar Proceedings
Strategic Datalinks
Strategic Profile Canada
The CISS Bulletin
Peacekeeping and International
 Relations (6 issues per year)
Canadian Defence Quarterly
(4 issues per year)

By Subscription:

The McNaughton Papers (The Canadian Journal of Strategic Studies — 2 issues per year)

The CISS is a non-profit, non-partisan voluntary organization which maintains an independent posture and does not advocate any particular interest.

For membership, seminar and publications information, please contact:

The Canadian Institute of Strategic Studies
Box 2321
2300 Yonge Street, Suite 402
Toronto, ON M4P 1E4
Tel: (416) 322-8128 Fax: (416) 322-8129
Email: ciss@inforamp.net
Internet: http://www.ciss.ca